lamiae JADOUAL

Optical emissions from Ni, Cr, Ti metal sprays

lamiae JADOUAL

Optical emissions from Ni, Cr, Ti metal sprays

ScienciaScripts

Imprint
Any brand names and product names mentioned in this book are subject to trademark, brand or patent protection and are trademarks or registered trademarks of their respective holders. The use of brand names, product names, common names, trade names, product descriptions etc. even without a particular marking in this work is in no way to be construed to mean that such names may be regarded as unrestricted in respect of trademark and brand protection legislation and could thus be used by anyone.

Cover image: www.ingimage.com

This book is a translation from the original published under ISBN 978-620-6-69081-8.

Publisher:
Sciencia Scripts
is a trademark of
Dodo Books Indian Ocean Ltd. and OmniScriptum S.R.L publishing group

120 High Road, East Finchley, London, N2 9ED, United Kingdom
Str. Armeneasca 28/1, office 1, Chisinau MD-2012, Republic of Moldova, Europe
Managing Directors: Ieva Konstantinova, Victoria Ursu
info@omniscriptum.com

Printed at: see last page
ISBN: 978-620-8-61208-5

Contents

My uncle
To my mother
To my aunt and her husband
To all my teachers

To all those who love me

Thank you

This work was carried out within the SIAM team (Spectroscopy and Atomic Imaging of Materials) of the Semlalia Faculty of Sciences at Cadi Ayyad University in Marrakech. I would like to express my warmest thanks to Mr **A. KADDOURI** for welcoming me to his laboratory and for guiding and directing the work on this thesis. Thanks to his great experience and perfect understanding, I was able to complete this work. His advice and criticism were invaluable. It is very difficult for me to express in these few lines all the gratitude and respect I owe him. I would like to express my sincere gratitude for his constant availability.

My sincere gratitude also goes to Mr. **A. OUTZOURITH**, Professor at the Semlalia Faculty of Sciences in Marrakech, for kindly chairing the jury. I would also like to thank him for making oxide samples available to us.

I would like to express my sincere thanks and deep gratitude to Professors **A. ARBAOUI, K. BOUABID and M. El AATMANI** for their kindness in judging and reporting this work. I am very grateful to them and I also thank them for their participation in this jury.

I would also like to express my gratitude to Professors **H. AOUCHICHE** and **H. OUGHADDOU** for agreeing to examine this work and for trouble they took in travelling from abroad to take part in the thesis jury. They have my deepest respect.

I greatly appreciated the availability and patience of Messrs **M. Ait El Fqih**, **A. El Boujlaidi**, **N. Rochdi**, **H. Afkir** and Mrs **R. Jordanie.** I would like to thank them for the fruitful discussions we had, their advice and also for the friendly atmosphere they created throughout the years of my thesis.

I would also like to thank all my friends who have given me moral support and have always helped and encouraged me in the completion of this work.

Finally, I would like to express my most sincere thanks to all the members of my family, in particular my uncle **AbdelKebir FARHANE** and my aunts **Mina GASSIM and Bouchra BELKZIZ** for the support and confidence they have shown me throughout this work. I am most grateful to my cousins **Malika, Imane** and **Ilhame** for their encouragement and humanity.

Summary

When a beam positive ions with a kinetic energy of a few keV interacts with a solid target, sputtering takes place and particles (electrons, neutral atoms, molecular ions, etc.) characteristic of the target are ejected. This type of interaction also causes the ejection of particles in electronically excited states that can be deactivated by the emission of electromagnetic radiation. The study of this luminous radiation has proved very interesting, as it has enabled us to elucidate the mechanisms by which excited species are formed in the vicinity of a surface, and to understand the processes by which these species are deactivated. Similarly, measuring the intensities and wavelengths of these radiations has made it possible to develop a technique for characterising the surface and depth of materials. This spectroscopic technique is known by the acronym SIPS (Sputter Induced Photon Spectroscopy) and is highly complementary to existing surface and interface analysis techniques.

This work is being carried out by the SIAM (Spectroscopy and Atomic Imaging of Materials) team at Cadi Ayyad University. It involves using the SIPS technique to study the luminescence of samples of Ni, Cr, Ti and the oxides NiO and Cr_2O_3. The targets are bombarded with Kr ions$^+$ of 5 keV at an angle of 65° to the normal and the spectral range explored extends from the near UV to the visible.

The studies we have carried out can be summarised as follows:

1) Demonstration of optical emissions from sprays of three metals: nickel, chromium and titanium bombarded in a vacuum (better than 10^{-7} torr).

2) Measurement of the optical spectra of these three metals in an oxygen atmosphere of 10^{-5} torr.

3) Measurement of the luminescence spectra of NiO and Cr_2O_3 oxides.

4) Identification of the spectral lines measured and comparison of their relative and absolute intensities in the absence and presence of oxygen.

5) Interpretation of the effect of oxygen on luminescence spectra.

6) Examination and application of the electron exchange model. Estimation of the electron affinity and band gap of oxides.

All the work presented has been the subject of five publications (3 already published and two in the process of being submitted) and around fifteen papers presented at international and national scientific events.

Key words: Ion bombardment, sputtering, luminescence spectra, metals and oxides, oxygen effect, electron exchange model.

الاشعاعات الضوئية الناتجة عن تفتت الأجسام الصلبة لنيكل، الكروم، التيتانيوم أكسيد النيكل وأكسيد الكروم الخاضعة للقصف الأيوني

ملخص

خلال التفاعل بين حزمة من الأيونات الموجبة ذات طاقة حركية تناهز بضع كيلوا إلكترون- فولط (keV) مع هدف صلب، يحدث تفتت وتتطاير جسيمات مميزة للهدف (إلكترونات، ذرات محايدة، أيونات، وجزيئات إلخ.....).

هذا النوع من التفاعل يؤدي أيضا إلى قذف بعض الدقائق المحفزة إلكترونيا تعود إلى حالتها الأصلية بإرسال إشعاعات كهرومغناطيسية.

إن دراسة هذه الإشعاعات الضوئية تكتسي أهمية بالغة لأنها تتيح معرفة آليات تكوُّن الأنواع المحفزة بجوار السطح وفهم كيفية عودتها إلى الحالة غير المحفزة. وكذلك فقد مكن قياس شدة وأطوال موجة هذه الإشعاعات من تطوير تقنية لدراسة خصائص السطح والعمق للمواد. هذه التقنية معروفة ب "SIPS" وهي مكملة للتقنيات المستعملة في دراسة السطوح والفواصل.

تم إنجاز هذا العمل من طرف فريق البحث "سيام" (SIAM) "المطيافية والتصوير الذري للمواد" بجامعة القاضي عياض. وتهدف هذه التقنية (SIPS) إلى دراسة إشعاعية لكل من النيكل Ni ، الكروم Cr، التيتانيوم Ti ، أكسيد النيكل NiO وكذا أكسيد الكروم Cr_2O_3، بفعل تعريضها إلى قصف أيوني بواسطة أيونات الكريبتون Kr^+ ذات طاقة حركية تصل إلى 5 كيلوا إلكترون- فولط (keV) ومن زاوية 65 درجة بالنسبة للمنظمي على السطح، يمتد المجال الطيفي من مجال الأشعة فوق البنفسجية القريبة إلى المجال المرئي.

يمكن تلخيص الدراسات التي قمنا بها كما يلي :

1- إبراز الانبعاثات الضوئية لكل من النيكل، الكروم والتيتانيوم عند تعريضها لقصف موضعي في الفراغ تحت ضغط أقل من 10^{-7} تور(torr).

2- قياس الأطياف الضوئية للمعادن الثلاث في محيط من الأوكسجين تحت ضغط يصل إلى 10^{-5} تور (torr).

3- قياس الأطياف الضوئية لكل من أكسيد النيكل NiO وأكسيد الكروم Cr_2O_3.

4- تحديد وقياس خطوط الطيف و مقارنة شداتها المطلقة والنسبية بغياب ووجود الأوكسجين.

5- تفسير تأثير الأوكسجين على الأطياف الإشعاعية.

6- فحص وتطبيق نموذج التبادل الإلكتروني واستخلاص كل من قيمة القدرة الإلكترونية و فجوة الطاقة (gap) للأكاسيد.

شكلت الأعمال المقدمة في هذا البحث موضوع خمس مقالات علمية (نشرت ثلاثة منها واثنتان في طور التقديم) وأزيد من خمسة عشر مشاركة في عدة مؤتمرات دولية ووطنية.

الكلمات الدالة : القصف الأيوني، دائرة التفتت، الأطيف الإشعاعية، المعادن والأكاسيد، تأثير الأوكسجين، نموذج التبادل الإلكتروني.

General introduction

This thesis describes a body of scientific work that we have carried out over the last four years. It focuses on the optical emissions measured during the sputtering of a solid surface subjected to the erosive action an ion beam.

It is well established that ion-solid interaction is a rapidly expanding field of research, since the various products of this interaction are often characteristic of the chemical composition of the target; they can then be used as the basis for techniques to study and characterise the target. Examples of such techniques include SIMS (Secondary Ion Mass Spectrometry) [1], SNMS (Sputtered Neutral Mass Spectroscopy) [2], RBS (Rutherford Backscattering Spectroscopy) [3] ISS (Ion Scattering Spectroscopy) [4], GD-OES (Glow Discharge Optical Emission Spectroscopy) [5], PIXE (Proton Induced X-Ray Emission) [6], etc. This type of interaction can also give rise to the emission of electromagnetic radiation resulting from the deactivation of particles formed during sputtering into excited states, and this technique is often referred to by the acronym SIPS (Sputter-Induced Photon Spectroscopy) [7,8]. Based on the detection of light emitted during ion-solid interaction, this technique consists of identifying, by measuring the intensities and wavelengths of the optical emissions, the particles ejected in excited electronic states during sputtering induced by energetic ion beam. The study of these optical emissions has proved very interesting both theoretically and practically.

From a fundamental point of view, the SIPS method is of potential interest, particularly for studying ion-surface interaction and understanding the behaviour of an excited or ionised atom in the vicinity a surface. In addition, the study of luminescence spectra has made it possible to develop models to explain the formation and de-excitation mechanism of excited species responsible for luminous radiation. It is also an interesting way of studying the phenomenon of adsorption; in fact, a great deal of work has been devoted to the analysis of surfaces in the presence of oxygen gas and its influence on light emissions [9,10]. Theoretical models have been proposed to explain this influence, and SIPS has been used to study the competition between radiative and non-radiative transitions of excited atoms sputtered near the surface. Similarly, the study of oxides by SIPS is very interesting because it has been possible to determine or confirm the values of the gap and the electronic affinity of the oxide [11,12].

From a practical point of view, examining the luminescence spectra enables us to determine the nature of the elements ejected and therefore the chemical composition of the bombarded sample. The spectral range explored generally extends from the near-UV to the visible (190 to 650 nm), enabling virtually all elements to be studied and identified. This technique is also a powerful tool for detecting impurities in the surface or deep layers of the samples analysed. SIPS has also been used successfully to study the diffusion of elements through nanometric layers [13].

The SIPS method is a related route to existing methods such : secondary ion mass spectrometry (SIMS), Auger electron spectroscopy (AES), X-ray induced photoelectron spectrometry (XPS) and laser ionisation mass analysis (LIMA). These surface analysis and profiling techniques provide very high precision on the composition of surface layers and interfaces. To obtain the maximum amount of information, especially for complex systems, it is often useful to combine several analysis methods. In fact, recent equipment incorporates several analytical and/or study textural on the same pumping system. In particular, the SIPS technique is considered to be highly complementary to the SIMS (Secondary Ion Mass Spectrometry) technique, since these two methods use the same type of excitation probe, i.e. an ion beam. However, SIPS has the advantage of not being sensitive to space charge, since photons are detected. Because it is only slightly destructive (0.5 Â/s), this technique should enable qualitative and quantitative analysis of the surface layers a target, as with static SIMS. The erosive action of the incident beam makes it possible not only to follow the insertion profile of the various elements in a given matrix, but also continuously examine the growth of thin layers or deposits.

The study we carried out in this work concerns the use of the SIPS spectrometer for samples of Ni, Cr, Ti and NiO and Cr2O3 oxides bombarded by Kr^+ ion beam with a kinetic energy of 5 keV. This spectrometer, installed in the SIAM (Spectroscopy and Atomic Imaging of Materials) laboratory at Cadi Ayyad University, has been widely used for a large number of fundamental and practical studies [11-16]. The experiments we carried out on these samples were performed successively in a good vacuum (<10 7 torr) and then in an atmosphere of a chemically reactive gas, this case oxygen.

Our aim is firstly to highlight the luminescence spectra of the samples studied and then to examine the effect of oxygen on the optical radiation emitted. We also compared our experimental results with the electron exchange model between the ejected species and the surface. This model has been widely used and validated for metals other than those studied in the present work [8,11,12,17].

The present work is intended as a contribution to the study of the light emitted during the sputtering of the surface layers of a solid subjected to ion bombardment. We also believe that the results obtained constitute a spectroscopic database for the three elements studied, namely : Ni, Cr and Ti and the oxides NiO and Cr2O3.

This report is divided into four chapters and an appendix:

The first chapter is described in two parts (A and B). The first is devoted to the fundamental aspects of the sputtering of a solid target. The second presents a chronology of some of the work carried out on the light emitted during the sputtering of a solid surface.

The second chapter gives a detailed description of the SIPS spectrometer installed in our laboratory and used to carry out our experiments.

In the third chapter, we present and discuss the luminescence spectra obtained for nickel, chromium and NiO and Cr2O3 oxide targets. We describe the experimental conditions under which these spectra were obtained and show that the presence of oxygen during sputtering is a key parameter in the photon emission yield.

The final chapter is devoted to the experimental results obtained when bombarding a clean titanium target in the presence of oxygen (oxygenated titanium). The luminescence of this target also reveals a continuous radiation (continuum) that we will discuss.

Finally, we will present in the appendix a study that we had to carry out in parallel with the above. It concerns the angular distribution of the sputtering products of a copper sample and a binary copper-beryllium alloy.

Chapter 1

Spraying a solid surface

origin of the light emitted during spraying

Part A

Spraying a surface

surface

Introduction

The interaction of a beam ions with a kinetic energy of a few keV with a solid sample results in the rejection of particles of various kinds, often with the emission of electromagnetic radiation. In the first part, we will review the fundamental aspects of the phenomenon of the sputtering of a solid target by an ion beam, as well as the parameters that influence this phenomenon, such as the nature of the target, the stechiometry, the degree of purity of the surface, the presence of oxygen, the nature of the incident ions, their energy and angle of attack.

In the second part, we will present details of the light emitted during the bombardment of a solid by energetic ion beam. We will describe the radiative and non-radiative excitation and deactivation mechanisms that are at play during the sputtering of the surface layers of a solid. These mechanisms have been explained by models generally developed on basis of hypotheses concerning the origin of the formation of excited states. Nevertheless, the fundamental question, the answer to which enables us to decide between the different models, is the following: does radiative excited state originate outside or inside the target? In other words, are the observed optical emissions due to :

> To the excitation of particles in a thin layer above the solid?

> To neutral particles ejected from the surface in the ground state and electronically excited by the incident ion beam or by fast electrons, also ejected from the surface?

> Excitations induced by collisions between slow particles and fast particles ejected from the surface?

Or quite simply: atoms that are excited inside the solid and leave the surface in a metastable state with high translational kinetic energies, thereby escaping the non-radiative deactivations caused by the presence of the solid surface.

I. Spray phenomenon

I.1. Description

When an energetic incident ion collides with a target atom, it transfers part of its kinetic energy to the latter. This first colliding atom can also collide with other atoms in the target, and three collision regimes can be distinguished, as shown in figure 1-1 [18] :

> **The simple collision regime:** In this regime, sputtering is characterised by the small number of atoms set in motion when the projectile ion slows down. In this case, sputtering is induced by very short sequences of collisions that can lead to the ejection of atoms (Fig. 1-1a). A backscattered incident ion can transfer sufficient energy to a surface atom to overcome the surface potential barrier. The energies are less than 1 keV.

> **The linear cascade of collisions :** In this regime, the projectiles have an energy of a few keV. The target atoms have acquired sufficient energy to generate cascade collisions that cause the atoms to be ejected from the surface (Fig. 11b). In this case, the moving atoms only collide with the stationary atoms; this is the general case of the sputtering used in the techniques known by the acronyms SIMS [1,19] and SIPS [7,12]. The latter is the technique used in this work.

> **The thermal spike regime:** In this regime, the projectiles have an energy greater than one MeV. The number of atoms set in motion is very large and in this case the collisions are between moving atoms. A cascade is no longer a sequence of successive and independent collisions between atoms in motion and atoms at rest. The mean free path between two collisions involving two moving atoms becomes small enough to define a volume in which all

the atoms are in motion (Fig. 1-1c).

For the last two regimes, the primary recoil atoms are energetic enough to generate secondary recoils. Some atoms can approach the surface of the target and break through the surface barrier.

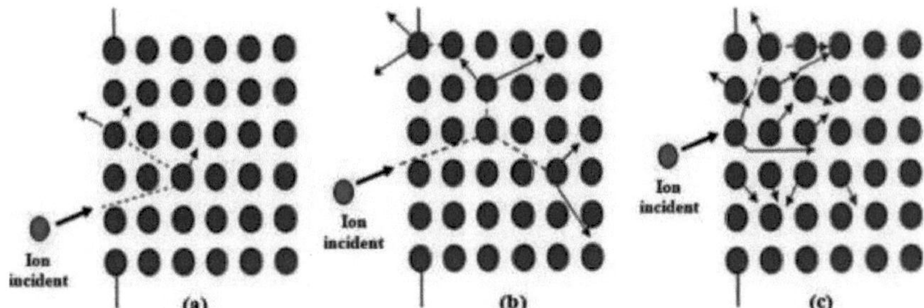

Figure 1- 1: Schematic representation of ion-surface interaction (a) Simple collisions (b) Linear cascades (c) Thermal spikes

I.2. Stopping power

Incident ions that have penetrated the solid eventually slow down and then stop in the solid at a depth that depends on the stopping power. The stopping power is the amount of kinetic energy lost by the ion per unit distance travelled. It can be broken down into three terms whose relative values depend, among other things, on the kinetic energy of the incident ion and its mass. These three terms are :

> Nuclear stopping power, which involves a direct collision between the nuclei of the incident ions and those of the target.

> The electronic stopping , the origin of which is due to a coupling between the incident ion and the electron cloud of the target atoms.

> The exchange stopping power (negligible), which involves an exchange of charge between the moving ion and the target particles.

The total stopping power a particle is expressed as :

r/v \ r/v / .

nucl elee

Figure 1-2 shows the variation in nuclear stopping power and electronic stopping power as a function of the projectile energy Cs on a Si target. This variation is simulated by the SRIM programme [20]. Nevertheless, the behaviour and orders of magnitude remain close for heavy ions and target atoms. Figure 1-2 shows three distinct zones:

> **Between 1 keV and 1 MeV**: Most of the energy loss is due to nuclear collisions. Collisions will therefore be quasi-elastic.

> **Between 1 MeV and 1 GeV**: The slowing down of the ion is mainly due to the interaction with the electrons in the target. The nuclear interaction with the target atoms becomes weak, which reduces the number of particles initially set in motion by the projectile. Conversely, the probability of ionising the target atoms increases.

> **After 1 GeV**: There is no longer any interaction with the target atoms and the electronic stopping power decreases as the energy increases. The energy range generally used for sputtering solids is the one for which the nuclear stopping power is predominant.

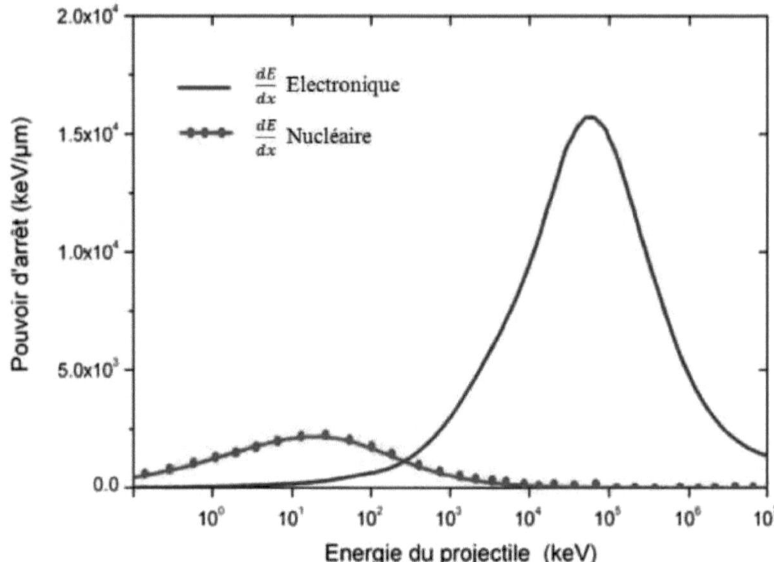

Figure 1.2: Loss of kinetic energy due to nuclear and electronic interactions as a function of the energy of the incident Cs particle on a Si target; after [21].

I.3. Spray yield

The main characteristic of the sputtering phenomenon is the sputtering yield, noted S. It is defined as the average number of particles (atoms or molecules) torn from the target per incident ion.

$$S = \frac{\text{Nombre de particules éjectées}}{\text{Nombre d'ions incidents}}$$

Number of particles ejected
Number of incident ions

This coefficient can be determined using a quartz microbalance by measuring the difference mass before and after bombardment. It can also be determined by measuring the depth of the cells using a microscope, or by directly measuring the quantity of material collected in a given solid angle.

Numerous parameters have a major influence on spraying efficiency, including :

> The energy of the incident ion, its mass and the angle of attack.

> The chemical state of the target, its crystallographic nature (crystalline or amorphous) and its temperature.

> The atomic number of the target.

> The degree of vacuum in which the bombardment takes place.

Most of these parameters are difficult to understand, which prevents a global analytical formalisation of the sputtering rate. However, for most amorphous or polycrystalline solids bombarded by ions of a few keV, S can be predicted satisfactorily by the linear cascade model.

Sigmund [22] gave an expression for S for a target containing atoms of mass M and atomic number Z bombarded under normal incidence by primary ions of mass Mp and atomic number z_p.

$$S_0 = \frac{0.042}{U} \alpha\left(\frac{M}{Mp}\right) s_n(E_p)$$

> 0.042 is expressed in A 2.

> $\alpha\left(\frac{M}{Mp}\right)$ is a function of the mass ratio obtained graphically.

> U is the binding energy in eV of the atom at the surface, which is generally taken to be equal to the sublimation energy.

> $s_n(E_p)$ the nuclear stopping power.

The variation in spraying efficiency with the angle of incidence of the beam is governed by the law :

$$S = S_0(\cos\theta)^{-5/3}$$

However, these calculations are not valid at very low energies E_p or for light ions because the energy transferred is insufficient to cascade collisions. Similarly, they do not apply at high energies or for heavy ions because the energy transferred produces a branched cascade in which virtually all the atoms in the region are set in motion. Finally, they are of no importance for large angles of incidence where a large proportion of the primary ions are backscattered.

II. Influence of bombardment parameters

The sputtering yield S, also known as erosion rate, depends the nature of the target, the nature of the incident ion and the attack conditions. The collision cascade is highly dependent on the ion-target pair and, depending on the case, the cascade may be very extensive or restricted.

II.1. . The energy of the incident ions

The sputtering efficiency depends on the energy of the incident ions. Figure 1-3 illustrates this variation for copper, molybdenum, nickel, iron and tungsten targets bombarded at normal incidence by N_2^+ and N^+ ions [23]. The maximum sputtering efficiency is reached at 5 keV and above.

When we look at the sputtering rate a target as a function of the energy of the incident ion, we see that there is an energy E_s (threshold energy) below which no atom can be ejected. There are three energy ranges the incident ion (Fig. 1-4) that can be described as follows [24] :

> **Region I**: Below the sputtering threshold ($E<E_s$); the ion-substrate interaction leads essentially to electronic exchanges. In this case, no atoms are ejected from the target.

> **Region II**: For a value equal to or greater than the sputtering threshold ($E > E_s$); the effects of the collision cascades begin to manifest themselves due to atomic displacements they produce. At this point, atoms can ejected from the target using the sputtering mechanisms described in the previous paragraph.

> **Region III**: Above the sputtering threshold ($E > E_s$); this is the active region of sputtering; there is a rapid increase in S, followed by a plateau of saturation which is rapidly reached if the masses of the ion-target pair are very different.

Saturation is also due to evaporation from the sample surface (thermal peak regime).

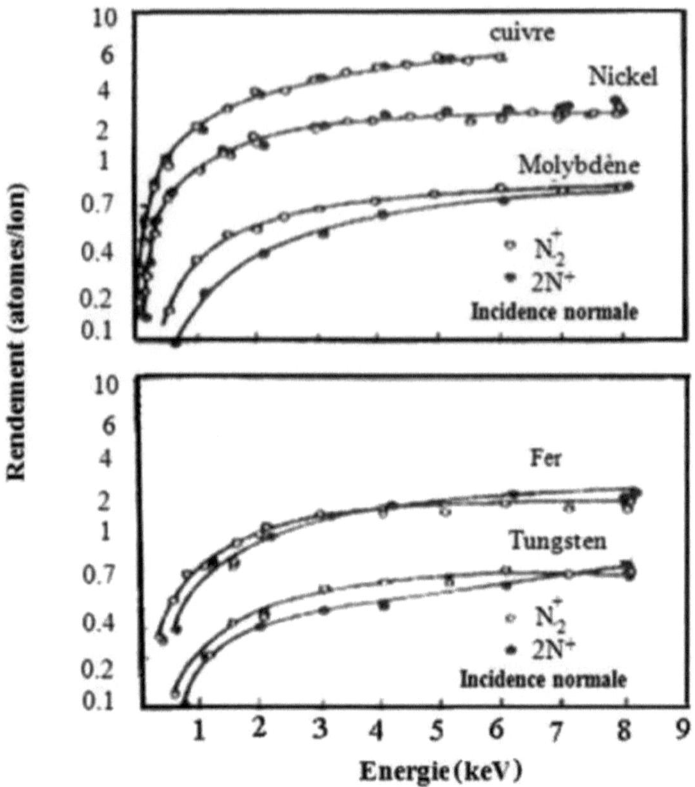

Figure 1-3: Variation in sputtering efficiency as a function of ion energy Primary for various targets; according to [23].

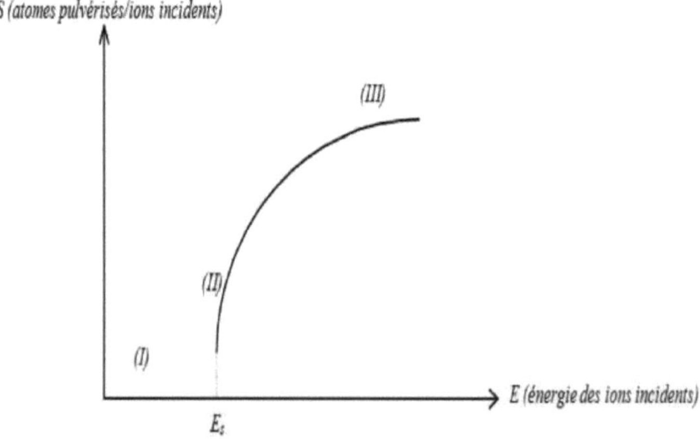

Figure 1-4: Variation in sputtering efficiency as a function incident ion energy; after [25].

II.2. . Angle of incidence

The angle θ which the direction of the incident ions makes with the normal to the surface has a very critical influence on the sputtering efficiency. In the case of copper, for example, S increases with the angle θ and reaches a maximum around 65 (Fig. 1-5); this means that the energy deposited as a result of the collision cascade in the outer layers and the number of atoms displaced in this region increases with the angle of incidence up to the value 65° . After this value, the sputtering efficiency falls and, in this case, the majority of the incident ions are backscattered by the atoms on the target surface.

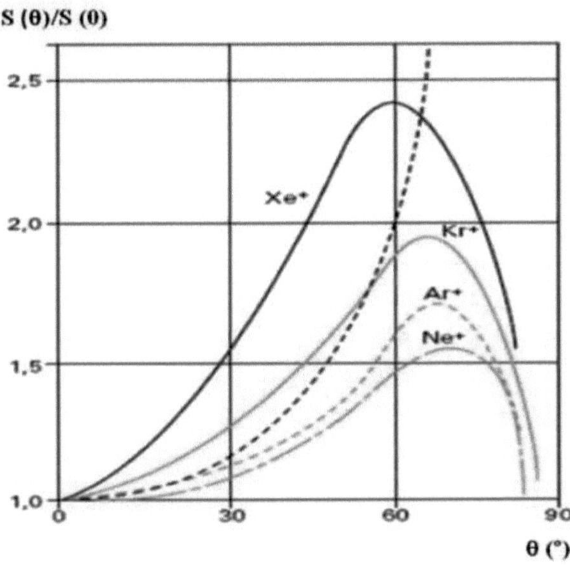

Figure 1-5: Variation of sputtering efficiency with angle of incidence for different primary beams of 1.05 keV energy on a copper target; from [26].

II.3. . Effect of residual atmosphere

The pressure of the residual gas in the vicinity of the surface of the bombarded sample has an enormous influence on the spraying yield. This effect becomes significant when chemically active compounds such as oxygen or caesium are introduced. Figure 16 shows the variation in S as a function of oxygen pressure for a silicon target bombarded with Ar^{+} argon ions with an energy of 5.5 keV [27]. The reduction observed reaches a factor of 4 when the surface is saturated with oxygen. Table 1-1 gives some sputtering yield values for metals as a function of oxygen partial pressure. It is clear that there is a decrease in the sputtering yield S when the oxygen pressure increases. The presence of oxygen leads to the formation by chemisorption of a layer containing metal atoms and oxygen, the nature and thickness of which vary from one metal to another. The metal-oxygen bond is stronger than that between the metal atoms and the rate of growth of the new layers formed is greater than that of the stripping of the surface atoms, which leads to a reduction in the sputtering yield.

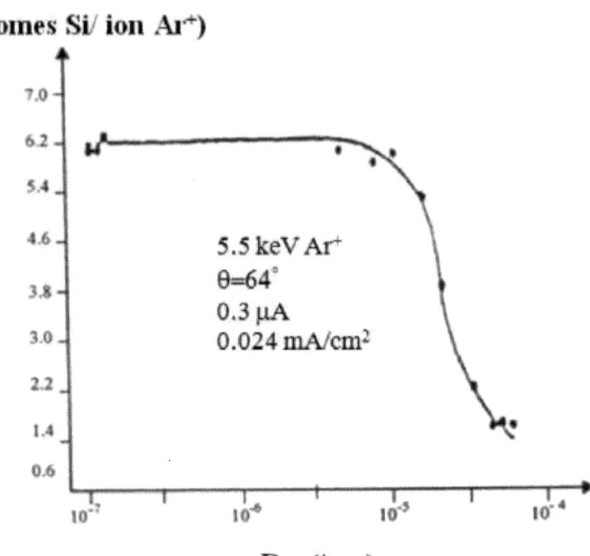

Figure 1-6: Variation in silicon sputtering yield as a function of oxygen partial pressure; after [27].

15 keV Ar+ on	*Cr*	*Zr*	*Your*	*Ti*	*Cu*
Clean target (10"^{7}Pa)	4.1	3.1	2.6	2.7	6.6
Poxy= 10 5 Pa	3.8				
Poxy= 5.10"^{5} Pa	1.8				
Poxy= 10-4 Pa	0.9	1.2	2.0		
Poxy= 10-3 Pa	0.8		1.4	1.0	4.0

Table 1-1: Variation in spraying efficiency for certain metals as a function of residual gas pressure (oxygen); after [28].

III. Spray products

III.1. 1. Composition of the total spray stream

The sputtering of a solid is accompanied by the ejection of particles of various kinds: electrons, atoms, ions, molecules, atomic or molecular associations. It is well established that some of these particles are ejected in excited electronic states [29]. They can then de-excite, outside the target, through radiative or non-radiative transitions. Furthermore, it is known that the presence of oxygen reduces the total sputtering yield (Tab. 1-1), but this reduction does not mean that all the sputtered species are affected in the same way. Betz and Husinnky [28] have shown that the partial pressure of oxygen in the vicinity of the sample acts differently depending on whether the sputtered particle is an ion, an excited atom, a neutral atom or a molecule.

Figure 1-7 illustrates the influence of the partial pressure oxygen on various sputter products a chromium target bombarded with Ar ions$^+$ of 15 keV [28]. For a pressure of less than 10 7 torr, the majority of the ejected particles are neutrals emitted in their ground state (curve A).

The relative proportions of particles emitted in different states of charge or atomic or molecular excitation do not exceed a few parts per hundred; the contribution of neutral molecular species remains negligible and decreases as their size increases. At oxygen partial pressures in excess of 10 6 torr, the composition of the spray stream changes considerably. The yield of neutral atomic species becomes negligible and the majority of secondary particles leave the surface in the form of molecules (AM curve). At the same time, the proportion of charged particles (atomic and molecular ions) increases and can reach 25% of the total flow (curve A^+) when the surface is saturated with oxygen. The yield of excited particles increases but their contribution, according to the authors, does not exceed 2% of the total flux (curve A*).

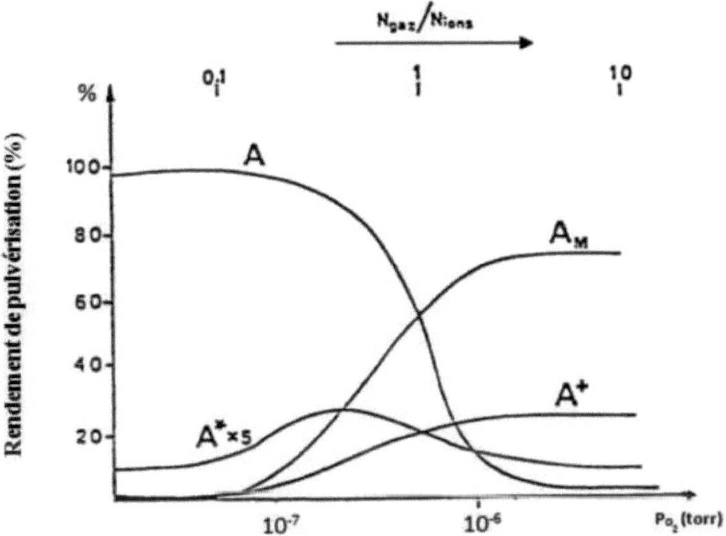

Partial oxygen pressure

Figure 1-7: Relative abundance of species ejected during bombardment of chromium by Ar ions+ of 15 keV, as a function of the oxygen partial pressure (A: neutral atomic atomic species,

AM: molecular species, A*: excited particles and A+ : ionised particles) [28]

III.2. 2. Energy distribution

The study of the energy distribution of the sprayed particles provides information on the processes responsible for particle emission. Figure 1-8 shows the energy spectrum of neutral particles emitted when Cu and Pt polycrystals are bombarded by Ar ions[+] at 3 keV [30]. The curve shows a maximum between 1 and 5 eV, the position of which depends on the nature of the target. Most of the particles are emitted with a kinetic energy of less than 10 eV.

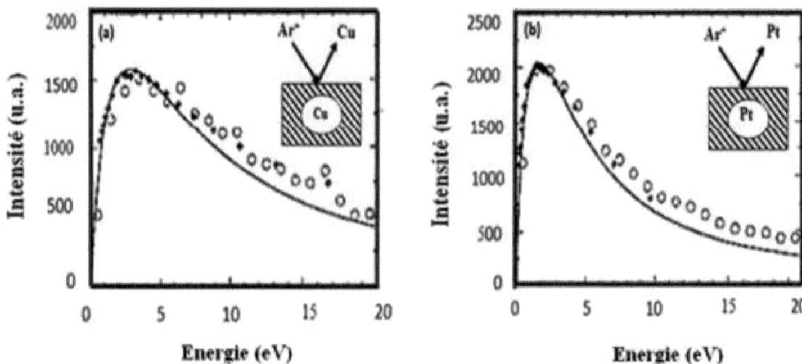

Figure 1-8: Energy distribution of copper and platinum sputtered under Ar bombardment+ 3 keV at 45 incidence [30].

Assuming an isotopic cascade and a plane barrier of height U0 for the surface potential, the energy spectrum of the sputtered atoms is governed by the law: calculations lead to the following expression [31] :

$$N(E) \propto \frac{E}{(E+U_0)^3}$$

U0 is the energy of the atoms at the surface and its value is usually taken, for polycrystalline targets, to be equal to that of the heat of sublimation. This formula has a maximum at $U_0/2$ and accounts for the asymptotic behaviour in $1/E^2$.

However, this distribution is no longer valid in the following cases:

> For medium-mass projectiles with energies in excess of 100 keV or heavy targets (Ag, Au) bombarded by heavy ions (Xe^+) of some keV.

> At low bombardment energies and/or in the case of bombardment by light ions (H^+ , D^+) for which we observe a shift in the maximum of the distribution and a significant contribution from particles emitted with a kinetic energy greater than 10 eV [32].

These observations show that the energy distribution depends on the mass and energy of the incident ions.

III.3. Angular distribution

The study of the angular distribution of the species emitted by a compound provides a detailed analysis of the preferential spraying by specifying, according to the experimental bombardment conditions, the preferential directions of ejection of the constituents. The angular distributions measured are, for the most part, described by a function of the type $\cos^n\theta$, with n between 1 and 2, θ is the angle made by the normal to the surface with the direction of the ejected particle. There are three angular distribution laws: cosine distribution, sub-cosine distribution and super-cosine distribution. The trajectory of sputtered atoms also depends on the ejection energy and it has been established that low-energy sputtered atoms have a sub-cosine distribution, while high-energy atoms have an over-cosine distribution [33 ,34].

Recently, we developed this study in the laboratory for products sputtered from a copper sample subjected to bombardment by Kr^+ ions with an energy of 5 keV. The ejected particles were collected on a sheet of Mylar and clearly showed an over-cosine lobe. Further experiments are required. This part will be presented in the appendix.

Part B

Origin of the light emitted when spraying a solid

I. Emission of photons by target particles

When ion beam interacts with surface atoms, a significant proportion of particles in the sputtered flux leave the target surface with a certain velocity in excited electronic states. These particles can de-excite at a distance D_{dex} such that $D_{dex} = V\_ . \tau$ or V is the average speed of movement of the excited particle in the direction perpendicular to the surface and τ the lifetime of its excited state. These particles, known as fast particles, are formed during violent inelastic binary collisions between the ions in the incident beam and the particles in the very first atomic layers of the solid. They are therefore ejected at high speed following violent inelastic collisions with the primary bombarding ion. Figure 1-9 illustrates the ejection and de-excitation of this type of particle.

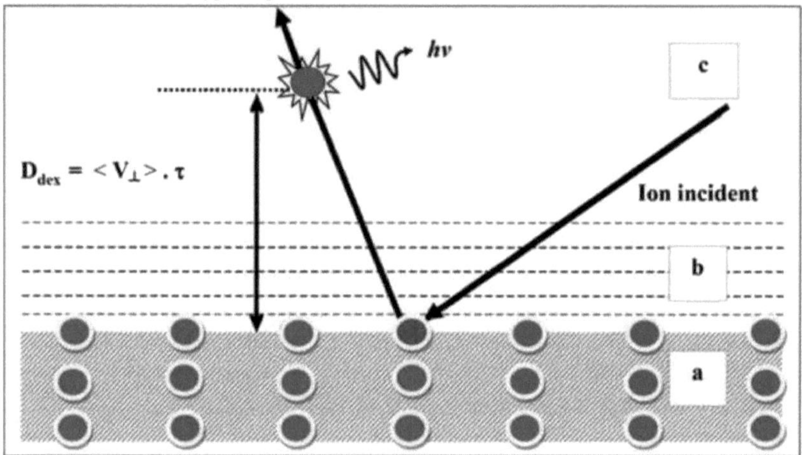

Figure 1-9: Schematic representation of the photon emission process by a surface atom hit by a fast ion.

a: metal, b: surface force field, c: free space

The average velocity of the emitting atoms was evaluated by Snoek et al [35] by measuring the extent of the luminescent region in front of the target using a study of the angular distribution of the emitted light. In the case of bombardment of a copper sample by Ar ions+ of 60 keV, the authors obtained for the Cu I resonance line (λ= 324.75 nm) a distance of between 0.25 and 0.35 mm which, given the lifetime of the radiative state (= 5ns), corresponds to a velocity of the order of 6.106 cm/s or a kinetic energy of 1 keV. This result was confirmed by the work of van der Weg and Bierma [36] who observed a displacement of 0.6 Â of the top of the same line. This shift, attributed to the Doppler effect, shows that the radiating particles are ejected with an energy of the order of keV. These observations led the authors to develop a model according to which the spectral lines are due to emissions from the de-excitation of fast particles. These particles, known as 'scattered' particles, leave the surface in excited states and emit photons at a macroscopic distance, s, (of the order of a mm) for which they can be considered isolated; their electronic states are no longer perturbed by the presence of the solid. In fact, it is well established that an excited or ionised particle located at a distance s very close to the surface is an unstable system that rapidly deactivates by non-radiative processes of the Auger type (de-excitation or neutralisation) or by resonance (neutralisation or ionisation). Figure 1-10 illustrates the configuration of potential curves of a system composed a metal and an excited atom located at a distance s from the surface. The metal is represented by a potential trough of depth εo filled with electrons up to the Fermi level of energy 8F and which is located at a distance equal to the work function of the metal φ below the vacuum energy (energy needed to move an electron away to infinity); the levels of the atom are also marked in relation to this energy. Ei being the ionisation energy of the free atom and E_x the energy of the excited electron. Four processes can occur close to the surface:

> **Resonant ionisation:** the excited electron is in a state with an energy (Ei-EA) lower than that the metal's Fermi level (Fig. 1-10a). This electron can tunnel to one of the unoccupied states of the metal's conduction band, giving rise to an ion which can eventually be neutralised by resonance (A*^ A+ + e "A).

> **Auger de-excitation:** the gap created in the excited atom is filled by an electron from the metal's conduction band (Fig. 1-10b). The energy acquired by the atom allows the excited electron to be ejected into the continuum (A* + e M ^ (A+ e A) ^ A+ e .Auger-). This process is governed by the energy condition: Ei- ε> Ei- EA or EA> ε where ε is the energy of the electron e M coming from the metal.

In the case where the particle involved is an ion, electron transfer can be effected by :

> **Resonant neutralisation**: for which the energy condition is written φ <E-E(A)<8 (Fig. 1-10c). This process can eventually lead to photon emission.

> **Auger neutralisation**: which gives rise to the rejection of an electron from the metal (Fig. 1-10d). In this case, the energy released by electron 1 is an energy between Ei-εo and Ei-φ, and must be greater than the energy ει of electron 2 in the metal. *£2* being between φ and εo which leads to at least the condition Ei>2 φ for electron 2 to be ejected from the metal.

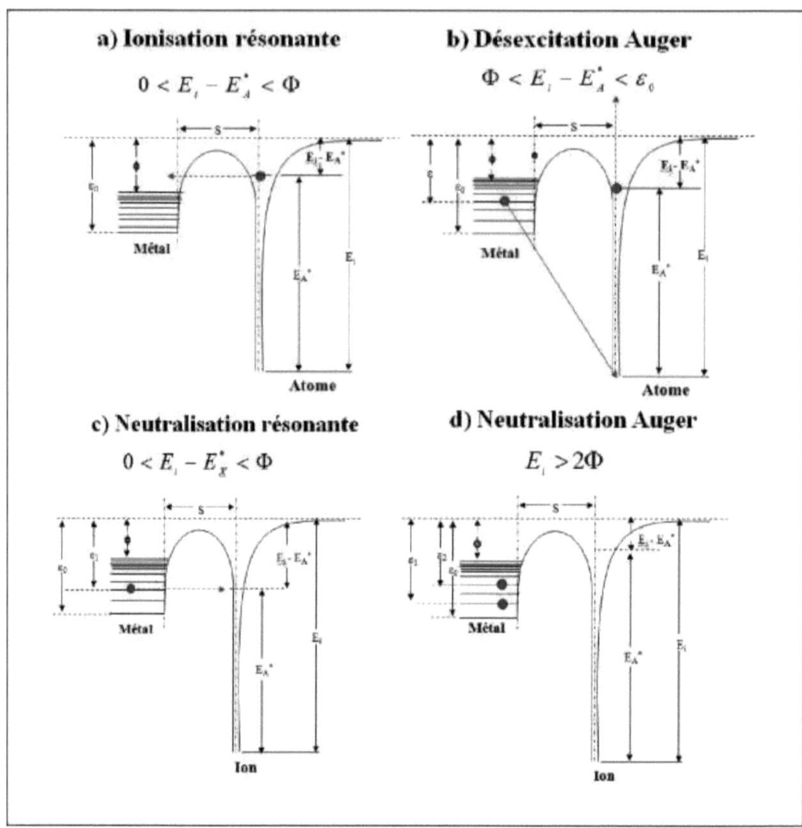

Figure 1-10: Electron exchange process between an excited atom or ion and a metal [37].

An excited particle very close to the surface tends to deactivate very quickly by exchanging electrons with the solid. Furthermore, the probability of an excited particle travelling a distance s without undergoing non-radiative de-excitation depends strongly on the normal component of its velocity according to the relation [38] :

$$P(s,v_\perp) \propto \exp\left[\frac{A}{av_\perp}\left(e^{-as}-1\right)\right]$$

which, for large distances s, is written as :

$$P(\infty,v_\perp) \propto \exp\left[-\frac{A}{av_\perp}\right]$$

For Auger and resonant processes, Terzic and Perovic [39] estimated A= IO^{14} -IO^{16} s ı is a= 2 A ı i.e. an average ratio 5.10^{13} Å/s. Adopting this value, Table 1-2 gives an estimate of the velocity and probability at two ejection energies of 10 eV and 1 keV for the three metals (Ti, Cr, Ni) that we studied (Chapters 3 and 4). We note in particular that, on average, the probability is 0.06 for slow particles (10 eV) and 0.75 for fast particles (1 keV). It is therefore clear that excited particles with a high velocity have a lower probability of undergoing the action of non-radiative processes and therefore contribute significantly to the radiation detected in the case of the clean targets studied.

Energies in eV	Ti		Cr		Ni	
	V (A/s)	P (%)	V (A/s)	P (%)	V (A/s)	P (%)
E= 10	$1,58.10^{13}$	0.042	$1,52.10^{13}$	0.037	$1,42.10^{13}$	0.029
E= 1000	$1,58.10^{14}$	0.728	$1,52.10^{14}$	0.719	$1,42.10^{14}$	0.703

Table 1-2: Velocity and probability ejection of particles with energies of 10 eV and 1 keV

II. Optical emissions from incident beam particles

When a target is bombarded by light ions (H^+ or He^+), rejection of particles from the target becomes less important, compared with that caused by heavy ions (Ar^+ , $Kr^+$the probability of backscattering of particles from the incident beam is appreciable [40].

The measurement of optical emissions reflects this phenomenon, and the spectra observed are dominated by lines associated with radiative emissions from excited backscattered particles (ions or neutrals). The interaction of incident particles with the nuclei can lead to an angular deviation in the trajectory of the projectile. The projectile, driven towards the surface, can leave with a high kinetic energy.

However, the interpretation of the mechanism for the formation of excited states remains very qualitative and it is assumed that the backscattered particles that are ionised or excited depend mainly on their velocity at the moment of rejection.

III. Influence of oxygen on optical emissions from metals

The introduction a chemically active gas modifies the composition of the surface and causes the formation a surface layer whose composition and thickness vary from one metal to another. Oxygen interacts with metal in three stages:

- Formation of a chemisorbed layer without modifying the crystalline order of the adsorbing metal.
- Penetration of the gas into the surface layers and surface reconstruction leading to the formation of a surface compound.
- Diffusion of the gas into the volume of the material and formation of a volume compound. As we have already pointed out, the presence of oxygen in the vicinity a metal surface during ion bombardment causes a decrease in the total sputtering yield, but also considerably modifies the proportions of the various species ejected from this surface. In particular, the presence of oxygen tends to reduce the number neutral atoms ejected in their ground state.

As for the light signal emanating from excited species, oxygen causes, for the majority of spectral lines, a more or less significant exaltation. Figure 1-11 shows the effect of oxygen and other gases on the neutral Al I line (396.15 nm) when aluminium is bombarded by Kr^+ ions with an energy of 5 keV. Hammoum et al [11] measured an increase in the intensity of the lines observed when iron was bombarded by Kr^+ ions of 5 keV in an oxygen atmosphere, as well as in the case of Fe2O3 and Fe3O4 oxides. This effect has also been observed by several authors for various targets bombarded in an oxygen atmosphere [8,41]. However, the presence of oxygen does not affect all the lines in the same way. Ait El Fqih et al [42] show that the presence of oxygen during aluminium bombardment causes an increase in the intensities of the Al I lines associated with neutral atoms. However, the intensity of certain ionic lines (Al II and Al III) decreases while others remain insensitive to the presence of this gas. Fournier et al [9] studied the influence of oxygen on the neutral Mg I and ionic Mg II lines. They observed an increase in the intensities of the neutral Mg I lines but a decrease in those of the ionic Mg II lines. The vast majority of these observations are probably governed by strong competition between the radiative and non-radiative transitions of the excited atoms and can be adequately explained within the framework of an electron exchange model between the ejected particle (neutral atom or ion) and the energy levels the surface. In particular, this model accounts for the observed exaltation, which appears to be due to a large contribution from the slow particles to the light signal. This model will be examined in detail in Chapters 3 and 4 for nickel, chromium and titanium targets bombarded by Kr ions .$^+$

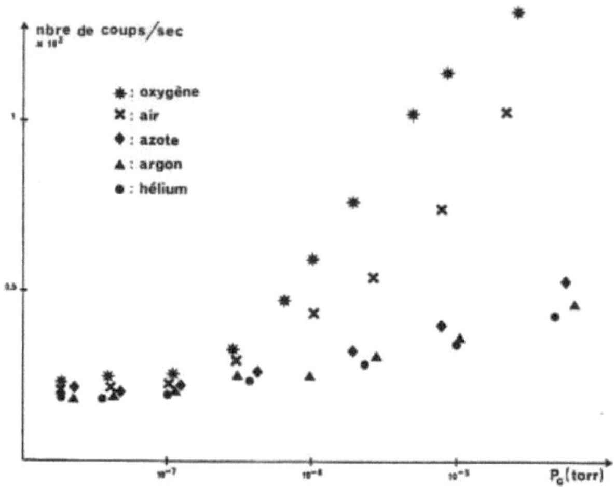

Figure 1-11: Variation in the intensity of the Al I line (396.15 nm) as a function of the nature and pressure PG of different gases during bombardment of an aluminium target by Kr ions+ of 5 KeV; after [40].

IV. Continuous radiation emission

Ion bombardment of certain targets gives rise to continuum radiation and there are several types continuum emission origin. The first type is continuum emitted by the solid itself (ionoluminescence). This type depends both on the composition of the surface of the bombarded target and also on the nature, energy and angle of incidence of the projectile. The second type is continuous radiation emitted by particles outside the target. Another type of continuous radiation can be envisaged when a solid target is subjected to ion bombardment and can be attributed to "Bremsstrahlung" braking radiation [43]. Sharodi et al [44] observed a continuum during bombardment of a beryllium target by He^+ ions at 15 keV, which they interpreted as resulting from electron-hole recombination. El boujlaidi et al [14] studied the luminescence spectra of six transition metals (Ti, V, Cr, Fe, Ni and Cu) during bombardment by Kr^+ ions at 5 keV. They observed that the continuum is much more pronounced for vanadium (3rd on 3d) than for titanium (2nd on 3d) and is totally absent in the case of chromium (5th on 3d), iron (6th on 3d), nickel (8th on 3d) and copper (3d is saturated). Consequently, they suggested that the continuum is present for elements whose 3d sublayer is less than half filled. The tendency of a metal target to give continuous radiation during bombardment by ion beam may therefore correlate closely with the number of electrons in the d layer. The origin of the continuous background often remains unexplained. Some hypotheses have attributed these backgrounds to molecular emissions of diatomic oxygen compounds. However, the explanation of this phenomenon has not been well detailed either by experimentalists or theorists. Understanding the origin of this radiation must necessarily involve a systematic study of all the elements.

To conclude this section, it can be argued that the light emitted during ion beam induced sputtering can have several origins. Figure 1-12 illustrates the different origins likely to be responsible for light emission. These include :

> Emissions in the form of discrete lines (hv_1, hv'_1) resulting from the deactivation of atoms or ions ejected in excited electronic states.

> Optical emissions in the form of a molecular band from excited molecules (hv_2).

> Radiation emitted by backscattered primary ions (hvs).

> Continuous radiation emitted by pulverised particles (hv_4) or from the solid itself (hv_5).

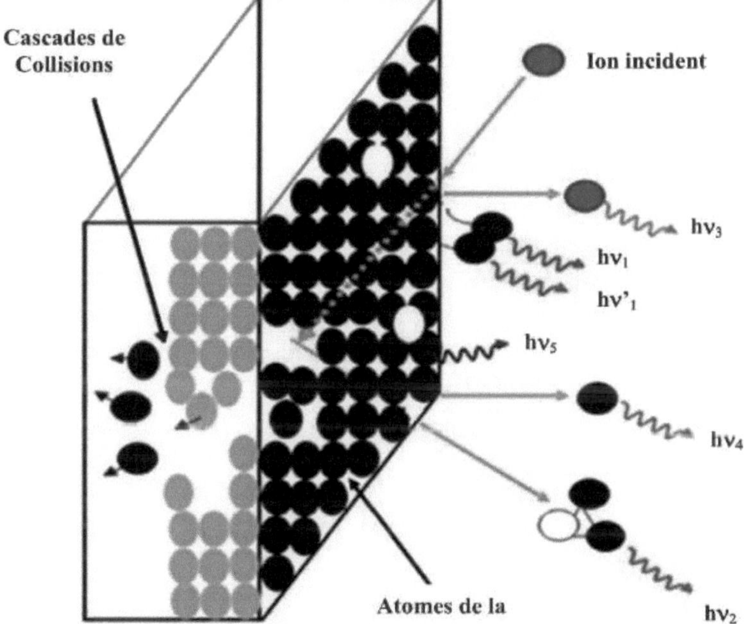

surface

Figure 1-12: Origin of the light emitted during spraying

Chapter 2

Experimental set-up

Introduction

The surface and solid characterisation method we use is known by the acronym SIPS (Sputter Induced Photon Spectroscopy). It is also known by other acronyms [7]: SCANIIR (Surface Composition by Analysis of Neutral and Ion Impact Rdiation), BLE (Bombardment Light Emission), IBSCA (Ion Beam Spectro Chemical Analysis) it was developed by researchers from different disciplines. In our SIAM "Spectroscopie & Imagerie Atomique des Matériaux" laboratory, we are developing it under the French acronym ASSO "Analyse de Surface par Spectroscopie Optique". This technique involves measuring the wavelengths and intensities of optical radiation emitted by atoms and/or molecules ejected in excited electronic states when energetic ions bombard a solid target. It enables the surface and in-depth composition of a solid to be determined qualitatively and quantitatively. The erosive action of the incident beam makes it possible to follow the insertion profile of the various elements in a given matrix. The ASSO apparatus is also a tool for studying ion-surface interaction and understanding the behaviour of an excited or ionised atom in the vicinity a surface.

This technique is considered complementary to the SIMS (Secondary Ion Mass Spectrometry) technique [45] since they both use the same excitation process, namely an ion gun.

I. Principle of the SIPS method

The principle of the SIPS method is relatively simple. A beam ions with a kinetic energy of a few keV strikes a solid target at a variable angle with a known angular aperture. As a result of this interaction, sputtering takes place and atoms or molecules in the target are ejected in excited electronic states. The light radiation emitted is analysed by a monochromator, which transmits the signal to a photomultiplier and then to an electronic counting chain. We observe photons from the far ultraviolet to the near infrared.

Surface analysis using the SIPS method, also known as ASSO, is therefore carried out by measuring the spectral lines emitted by the atoms on the surface. The advantages of this method include :

> The ability to identify the elements making up a solid sample.

> Determine the internal energy of the species sprayed.

> Elucidate the behaviour of an excited or ionised atom in the vicinity of a surface.

> The right resolution, since the spectral resolving power exceeds 300,000, which is better than 0.015Â at a wavelength of 5000Â.

> The solid angle for collecting photons is around 10 times greater than that for collecting material particles (secondary ions, for example).

Figure 2.1 illustrates different atomic emission processes when an ion beam strikes a solid. While interaction processes are numerous, the particles ejected from the surface are also very diverse: ions, neutral atoms, excited atoms, molecules and aggregates. Electrons can also be emitted from the surface or from particles ejected in excited states. Incident particles are partly backscattered. The spectral range used in this study extends from 190 to 650 nm and is easily accessible from an experimental point of view.

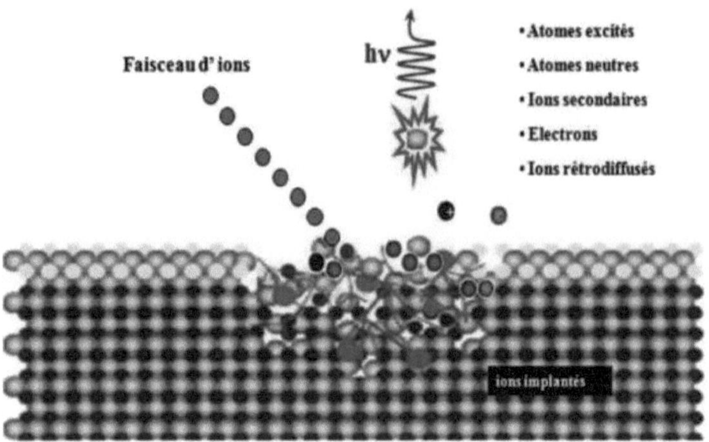

Figure 2-1: Principle of the SIPS method

II. SIPS equipment

The SIPS apparatus, as we have developed it in our laboratory under the acronym ASSO, is a modular spectrometer that enables the study of optical radiation emitted during sputtering induced an ion beam. This equipment is essentially made up of three removable parts that can be modified according to experimental requirements:

> An ion cannon.
> An enclosure fitted a sample holder.
> A system for detecting and analysing light emissions.

The first two parts are shown Figure 2-2 and the whole system be described in detail later. This spectrometer has been widely used to reveal luminescence spectra of several light and heavy metals [17,42]. It has also been used to study the interdiffusion elements across the interfaces of nanometre-thick layers (Co/AlO/Si) [13] and thin layers (C/W/Si) [46].

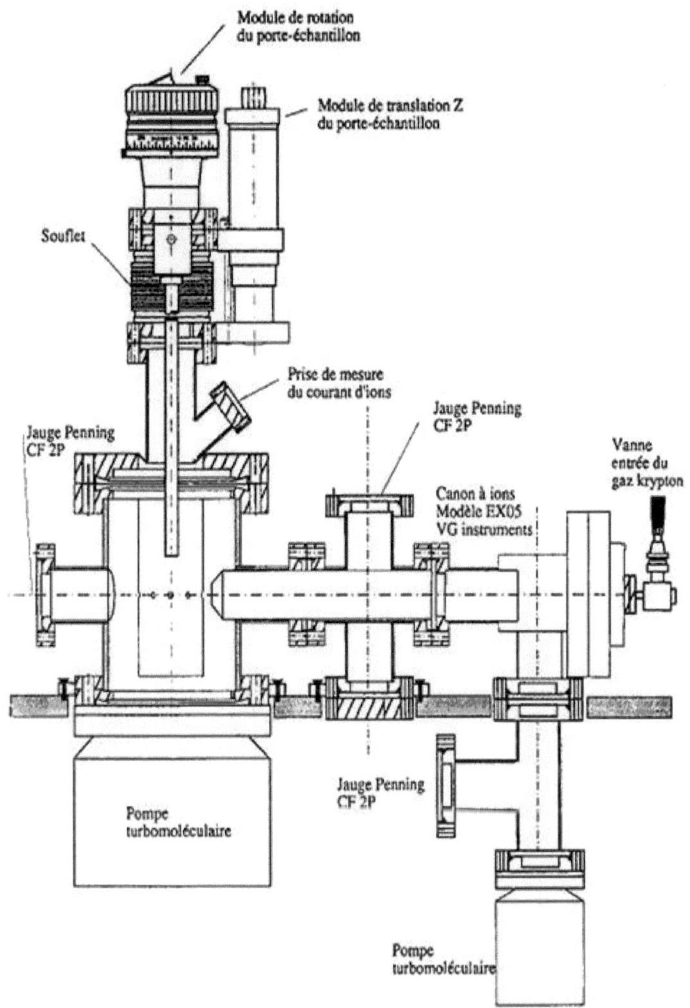

Figure 2-2: General diagram of the ion gun and sample chamber with pumping system

11.1. The ion cannon

11.1.1. Description

The ion gun, model EX05, consists of a source for producing ions by electronic impact and a system for extracting and focusing these ions. This gun can produce a beam with a kinetic energy of between 0.1 and 5 keV and a current of a few microamperes.

The ion source (Fig. 2-3), of the monoplasmatron type, consists of a stainless steel cylinder 127 mm long with an internal diameter of 35 mm. The anode is located 7 mm from the cathode and consists of a 0.5 mm diameter tantalum filament wound in a double spiral. The entire body of the source is connected to the high voltage and is electrically isolated from the rest of the device by a Teflon flange. The gun is pumped by two pumps connected in series, a 50 l/s Leybold Turbo molecular pump connected to a primary pump (Alcatel). The vacuum in

the source reaches 10 7 torr without the introduction of gas to be ionised.
The ion optics that control focusing are made up of two electrostatic lenses, each with three electrodes. The one closest to the source is used to condense and adjust the cross-section of the ion beam, while the other focuses it on the target to be bombarded. The diaphragm between these two lenses reduces the size of the beam to a diameter of <1mm. The deflection system is made up four plates forming two flat capacitors and allows the beam to be adjusted in two directions perpendicular to that of the beam.

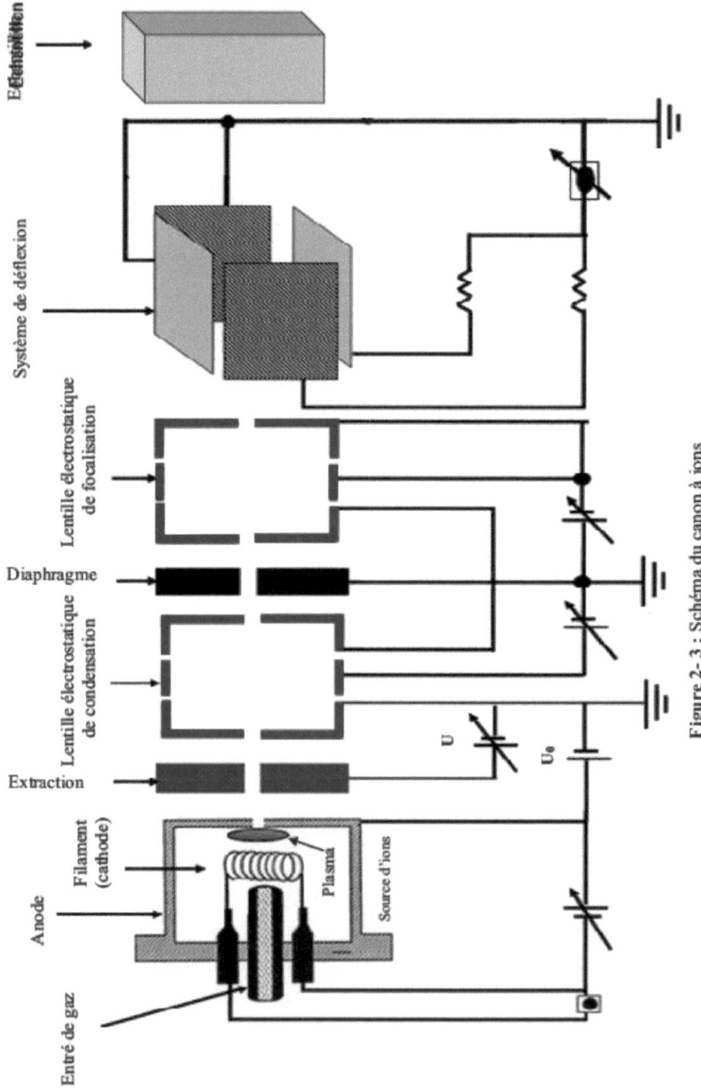

Figure 2- 3: Diagram of the ion cannon

11.1.2. Operating protocol

The procedure for producing Kr^+ ions in the electron impact source consists of :

> Introduce the gas to be ionised (krypton) into the source's collision chamber using a valve (Varian variable Leak Valve) at a pressure of between 10 6 and 10 5 torr. A pressure of 5.10 6 torr is often used in our experiments in order to extend the life of the source filament.

> Switch on the source (400X power supply) and wait for the filament current to stabilise at 2.5 A.

> Gradually increase the electron emission current from the filament until its intensity reaches a value of 2.5 mA. The electrons emitted by the incandescent filament collide with the Kr atoms in the gas, ionising it as follows:

$$Kr + e^-_{filament}(E_c) \rightarrow Kr^+ + e^-_G + e^-_{filament}(E_c - I)$$

Ec is the kinetic energy of the incident electrons at the moment of impact, I is the ionisation potential of the Kr gas, which is 13.99 eV. The electrons eG accelerated by the potential gradient are in turn likely to ionise other Kr atoms.

> Apply the acceleration voltage to the ions formed in the source. This voltage can be varied from 0 to 1 kV by a continuous potentiometer and from 1 to 5 kV by another potentiometer in steps of 0.5 kV. We worked with Kr^+ ions which we accelerated to a voltage of 5 kV. The krypton gas (N48) to be ionised is 99.998% pure.

> Adjust the diameter, focal point and direction of the beam. The three adjustments are made independently of each other and are designed to optimise the beam to give a diameter of less than 1 mm and an intensity of the order of pA.

11.2. The enclosure and sample holder

11.2.1. The enclosure

The sample chamber (Fig. 2-4) is a steel cylinder 125 mm in diameter and 143 mm long. It has eight circular side openings, four of which are 26 mm in diameter, two are 35 mm in diameter, one is 39.5 mm in diameter and the last is 41.5 mm in diameter, which is connected directly to a 200 l/s turbomolecular pump (ALCATEL, model Turbo V200A). This chamber is equipped with a valve (micro Hoke mite) connected to a 2 mm diameter capillary, the end of which is 2 cm from the sample to be bombarded. This valve allows reactive gases to be introduced into the chamber. In our experiments, we used oxygen (N48) of 99.998% purity.

11.2.2. The sample holder

The sample holder is made of aluminium and machined to minimise the secondary effects of sputtering, i.e. the bombardment of the sample by the sputtered particles and the secondary effects of ion beam rebound.

This specimen holder (Fig. 2-5) contains four compartments to hold four specimens, each measuring 10 mm x 10 mm; this enables alloys and their constituent metals to be analysed, for example. The samples are fixed to the back of the sample holder by means of a counter-plate. The latter is isolated from the rod attached to the rotation module by a Teflon clamping ring, and the volume trapped between the rod and the ring is pumped through a hole drilled in the sample holder. The sample holder is electrically connected to a pico-ammeter, which measures the ion current on itself. A micrometric stage associated with a rotation system is used to support the sample holder. To study the dependence of the signal on the angle of attack, the sample holder is rotated about the vertical axis. Translation along this axis allows the sample holder rod to be moved and the sample to be analysed to be selected.

The SIPS spectrometer can also be used to analyse powders using sample holders that we have designed and built in stainless steel. These sample holders can be fitted directly onto the support of the main sample holder (Fig. 2-6). A sufficient quantity of powder can be placed and compacted by hand using a rod also made of stainless steel, or compacted using a mechanical press.

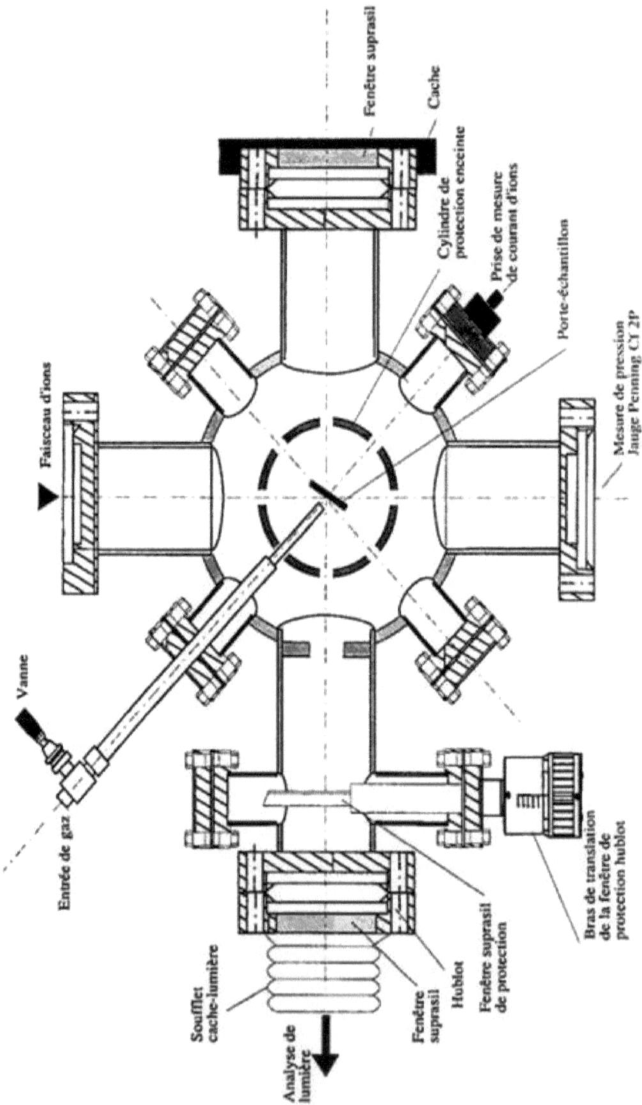

Figure 2-4: Schematic view of the sample enclosure (side view)

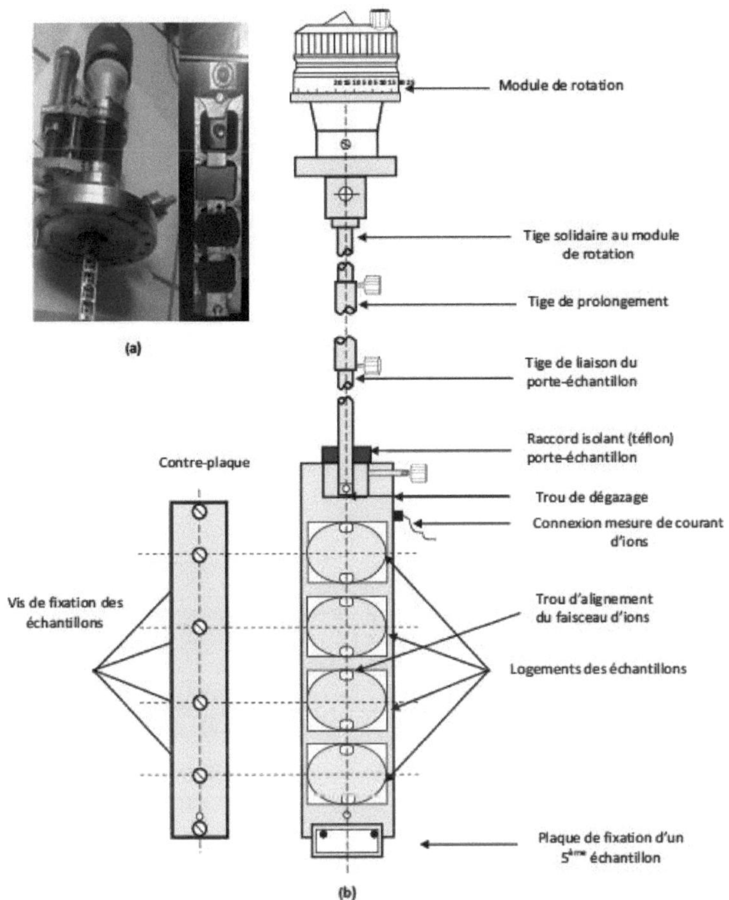

Figure 2-5: (a) Photo of sample holder with samples in place
(b) Diagram of sample holder+ mounting plate

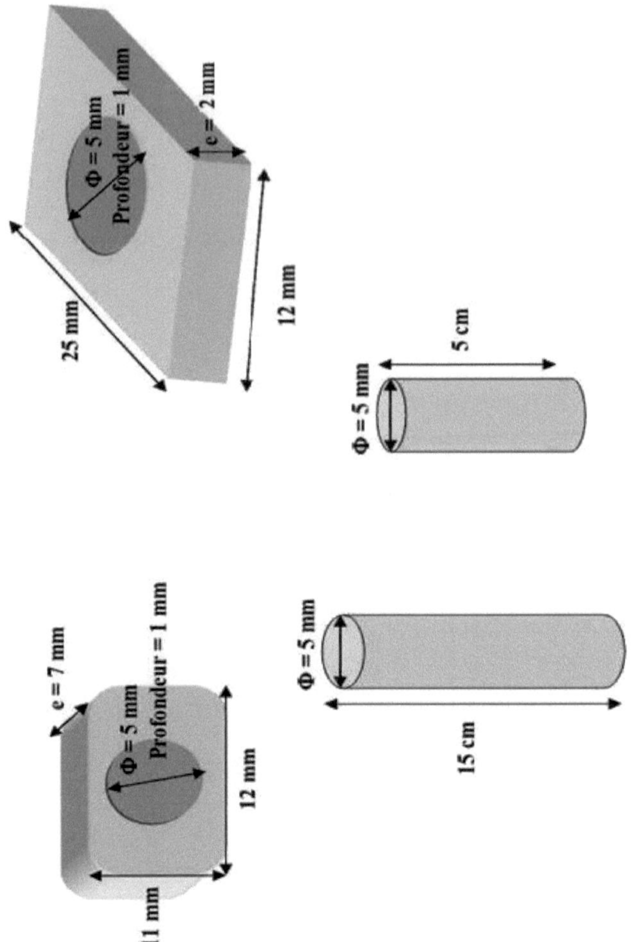

Figure 2-6: Diagram of the two sample holders with their rods for compacting the powders

11.2.3. Preparation of samples

In our experiments we cleaned the surface of our samples using the following procedures:

a- Polishing

Polishing is an important stage, the aim being to obtain a smooth, virtually flat surface. There are several types of polishing: mechanical polishing, electrochemical polishing and ionic polishing. In our experiments we used mechanical (manual) polishing with discs of different materials rotating at high speed, with or without polishing paste. The steps involved in polishing the surface of the sample are :

> **Step 1**: use silicon carbide SiC abrasive grains of variable size, shape and orientation; this enables material to be removed on a microscopic scale by abrasion.

> **Stage 2**: consists of polishing the surface with diamond pastes of decreasing particle size (6 μm, 3μm and hum) on rotating felt belts using a lubricant that dilutes the diamond paste of

the PS4/HV type.

> **Step 3**: use mechanical micro-polishing with alumina grains of 5 to 0.5 μm diluted with distilled water and using a SAT D1 reference disc.

There is a whole range of silicon carbide abrasive grains in different sizes, all working on the same principle. The different grain sizes for coated abrasive products are based on the ISO standard:

> **Coarse :**

P 80 - 60 - 50 - 40 - 36 - 24 - 20 - 16 - 12

> **Average :**

P 280 -240 - 220 - 180 - 150 - 120 - 100

> **Fine :**

P 600 - 500 - 400 - 360 - 320

> **Very fine :**

P 1500 - 1200 - 1000 - 800

b- Ultrasonic bath

In this type of process, the material is immersed in a liquid in which ultrasonic waves are propagated. Water is generally used as the propagation medium, but an organic solvent or an aqueous solution can also be used depending on the desired application. The ultrasonic wave creates zones of compression and depression in the propagation medium, leading to the formation of micro-bubbles, known as cavitation bubbles, which contain water vapour and dissolved gases. The implosion of these bubbles produces a very high velocity jet that physically cleans the surface of the material. For our samples, cleaning was carried out a pure acetone bath for 30 minutes. The sample is then dried with a hair dryer, care to remove all traces of solvent. The sample is then ready to be introduced into the apparatus and evacuated.

11.3. Light analysis and detection system

The light emitted by the particles ejected when the sample is bombarded is observed through a 10 mm thick suprasil window mounted on a translation arm that also protects the porthole from projections. The size of this window, twice the diameter of the opening in the protective cylinder, means that a new transparent zone can be exposed without having to break the vacuum.

Figure 2-7 shows the transmission curve of the 10 mm thick suprasil window compared with that of quartz of the same thickness. This window provides very appreciable transmission, without deformation, over a wavelength range extending from 1700 Â to 10000 Â. At the outlet of the chamber, a biconvex lens of the same thickness, with a diameter of 40 mm and a focal length f = 140 mm, is placed at a distance 2f from the sample and the monochromator input slit. The monochromator comprises a holographic grating, two concave spherical mirrors and two controlled slits. The entrance and exit slits are adjustable in height by a step of 2, 5, 10 and 20 mm and in width from 0 to 2 mm by a micrometric screw allowing a step of 10 μm. The light beam is protected from stray radiation by two black bellows on either side of the lens. A photomultiplier (PM) connected to a counting chain and to the Spectralink, which controls the PM power supply and data acquisition. A microcomputer equipped with an acquisition card to control the stepper motor driving the array. It processes the data received using the PRISM programme developed by Jobin-Yvon. The data is then stored in ISA-type files that can be converted to ASCII format by the same programme. The various devices in the detection and data acquisition chain are shown in Figure 2-8.

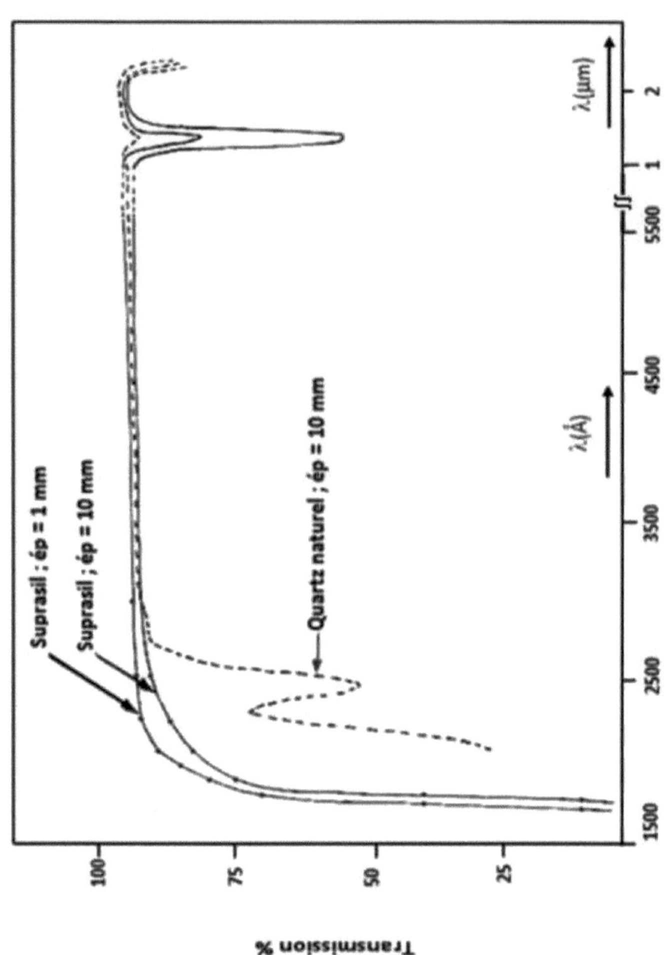

Figure 2-7: Transmission of suprasil and quartz windows [40].

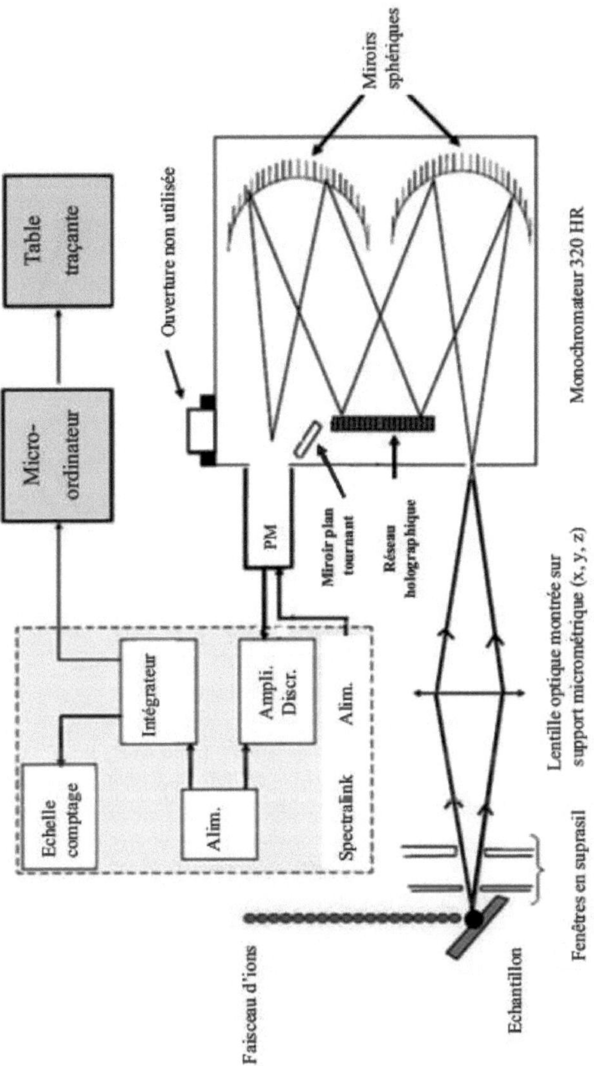

Figure 2-8: Light beam path and detection and acquisition chain

11.3.1. The monochromator

The spectrograph/monochromator we used is a Jobin-Yvon HR-320 of the Czerny-Turner type with a focal length of 320 mm and equipped with a holographic grating and two concave spherical mirrors (Fig. 2-9). Dispersion or diffraction of the light beam can only be used if the light is collimated, i.e. if all the light rays are parallel (or practically parallel). In practice, light beams are almost always divergent and it is necessary to use a collimator to make the rays parallel. Generally speaking, monochromators use collimators that work in reflection

(and not in transmission) so as not to introduce any parasitic dispersion of the light.
The light beam passing through the entrance slit E is reflected by the collimating mirror M1, which makes it parallel and then directs it towards the grating R. The grating diffracts the light and sends a spectrum to M2, which focuses an image of the entrance slit on the exit slit S. The wavelength is scanned in the plane of the output slit by rotating the grating. The grating is rotated by a stepper motor controlled by a microprocessor, so that the wavelength can be scanned linearly. The wavelengths are displayed directly in angstroms on a digital counter.

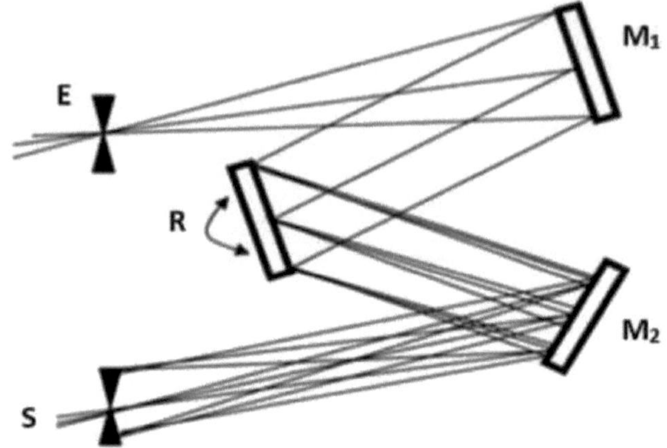

Figure 2-9: Optical diagram of the monochromator
*E: Input slit; **R:** Network; **M1** and M2: Mirrors; **S:** Output slit*

11.3.2. The photomultiplier

The photomultiplier (PM) is a device that converts light into an electrical signal that can then be processed electronically. It consists of a photocathode, several dynodes and an anode plate. When light enters the tube and strikes the photocathode, electrons are pulled out by the photoelectric effect and emitted into the vacuum space of the tube. These electrons are attracted to secondary electrodes (dynodes) raised to a higher potential. The electrons accumulate at the output of the anode, amplifying the input signal (Fig. 2-10).

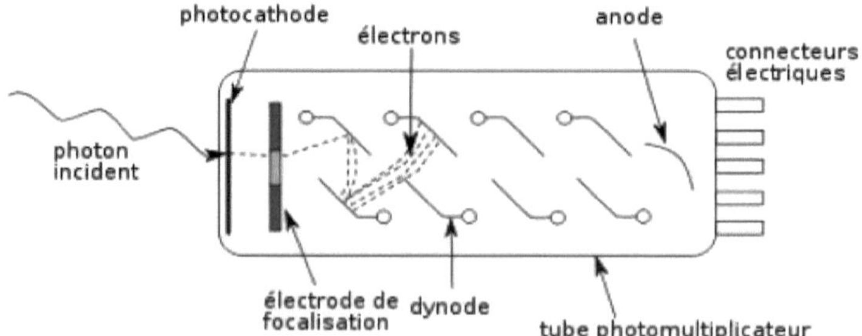

Figure 2-10: Photomultiplier operating principle
The Hamamatsu 4220 P photomultiplier we used is a photon-counting PM with a fused silica window and transparent photocathode. Its spectral range extends from 185 to 710 nm, with a maximum response at 420.0 nm (Fig. 2-11a). The PM has 10 dynodes (stages) with high collection and secondary emission coefficients, arranged to generate an electronic avalanche

emission process. Figure 2-11b shows the variation of the parasitic current -the dark current- with wavelength. This variation is due to thermoelectronic emission and this current is very small for the wavelength range we have explored. The gain of the PM is of the order of 10^7 when the voltage applied between the anode and cathode is 900 V (Fig. 2-11c), the voltage for which the signal-to-noise ratio is optimum.

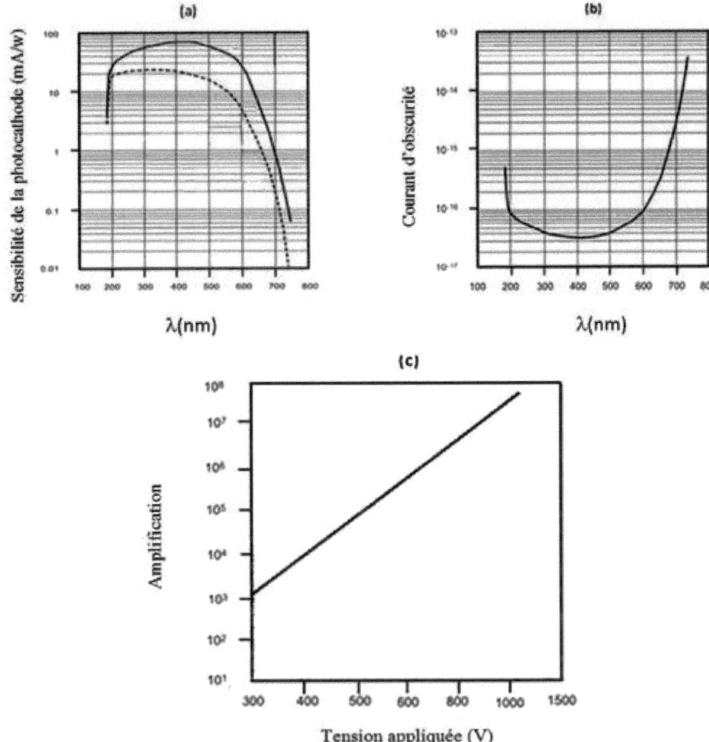

Figure 2-11: Characteristics of the PM used

(a) Photocathode sensitivity as a function of wavelength
(b) Variation of dark current as a function of wavelength
(c) Gain variation as a function of applied voltage

11.3.3. Spectralink

The spectralink is made up of a set of modules: stepper motor power supply, high voltage for the PM, data acquisition and recording, input/output, multi-channel control, multi-channel memory, multi-channel analogue and interface/computer module. Each spectralink module is equipped with its own microprocessor and instruction menu. Each module also has its own address number (4 bits) selected according to configuration.

Internal data exchanges with all modules are carried out via an 8-bit bus. External computers can access this bus using the interface module (INT). This allows full bidirectional communication in either parallel or serial mode.

a- Parallel mode

This mode manages :

> 8-bit data transmission and reception ;
> transmission and reception of control signals ;
> receiving addresses ;
> reformatting signals sent and received ;
> management of all signals by a microprocessor.

b- Serial mode

This mode allows :

> isolation of transmitted and received signals using optical couplers;
> reshaping transmitted and received signals ;
> management of all signals using a microprocessor;
> RS 232 C standard signal processing with switchable transmission speed.

11.3.4. The I.S.A card and "PRISM" software

a- The I.S.A. card

The I.S.A. (Industry Standard Architecture) acquisition card is an accessory used in a microcomputer to automate measurements. This automation requires two main functions:

> Acquiring data from the outside world is the role of inputs.
> The outputs are used to control the computer's actions.

The I.S.A. card is installed in the IBM PC bus (or 100% compatible). It is managed by the spectralink. This card is installed in the computer in the same way as any other card, you just need to follow the instructions provided by the computer to access the procedures. The I.S.A. interface card has a predefined 300 hex or 768 decimal address port. The flat link cable is attached between the rear of the spectralink's internal board and the connector on the I.S.A. board. This cable should be oriented so that the colour on the top of the spectralink faces the bottom of the computer, i.e. the cable should have a twist.

b- "PRISM" software

The reception and processing of experimental signal data are managed by the microcomputer using the PRISM Jobin-Yvon programme. This software is managed by a menu and its execution is chosen by selecting the desired function with the cursor and validating by pressing the SHIFT and F10 keys simultaneously. The main menu appears as soon as you launch the PRISM software, which offers a host of useful functions for starting up data acquisition and processing.

The following options are available in this menu:

> **Acquisition:** program controls the stepper motor that moves the monochromator grating and acquires the signal received by the PM. The acquisition function only works if the ISA card is installed and the spectralink is switched on.

> **Information:** This section contains information on the various spectral acquisition parameters and data manipulation in the files.

> **Manipulation:** This section allows you to view, plot and mathematically manipulate the acquired spectra. The manipulations will only work if the PRISM card is switched on.

> **System utilities:** This program is independent of the spectralink and the ISA card.
It makes it easy to use the sections, manage start-up and automatic back-up.

> **ISA-ASCII conversion:** This program converts ISA files into ASCII format for use with other software such as Origine, Excel, etc.

> **3D VIEW:** This option allows spectra to be viewed in three dimensions.

> **RETURN TO OS**Exits the software.

Chapter 3

Optical emissions from sprays of nickel, chromium and NiO and Cr_2O_3 oxides

Part 1

Optical emissions from nickel and its oxide NiO

Introduction

In this chapter, we present the experimental results obtained by bombarding samples of nickel, chromium and their oxides NiO and Cr2O3 with 5 keV Kr^+ ion beam. The Ni sample is solid, while the NiO, Cr and Cr2O3 samples are in powder form. In the first stage, we carried out the experiments on the four samples in a vacuum of better than 10^{-7} torr. In the second stage, we bombarded the Ni and Cr samples in an oxygen atmosphere of 10^{-5} torr. All the luminescence spectra were recorded under the same experimental bombardment and detection conditions.

The luminescence spectra we have obtained consist of a series of very fine lines, and each spectrum is recorded several times in order to check the reproducibility of the spectral lines and their intensities.

In the presence oxygen, we measured an increase in the intensities of the spectral lines, which is discussed in the context of the electron exchange model between the ejected particle and the energy levels the solid. Using this model, which assumes that the presence of oxygen modifies the energy band structure of the metal, we examine the competition between the radiative and non-radiative transitions of the excited species formed in the vicinity of the bombarded surface. This allows us to understand the variations observed in the photon yield.

It should be noted that these results have been the subject of several oral communications and two publications which we have cited in the references [12,47].

I. Physical and chemical properties of nickel

Nickel is a silvery-white metal with a fibrous fracture (Fig. 3-1), atomic number 28, average molar mass 58.7 g.mol^{-1}, electronic configuration [Ar] $3d^8 4s^2$ with $^3F_4$as the ground state spectral term.

Nickel is one of the most abundant elements in the universe, but is only found in small quantities on Earth. It is hard and ferromagnetic up to 355° C (Curie point) and is also a good conductor of heat and electricity. Its magnetic and chemical properties are similar to those of iron and cobalt. Its high level of resistance explains its use in the chemical industry. Its exceptional resistance to corrosion is a remarkable quality, hence its use in nickel plating.

Nickel is used in a wide variety of applications (dentistry, coins, electric guitar strings, boat hulls, electroplating or nickel plating, steelmaking: production of stainless and heat-resistant steels with very low cobalt content for nuclear applications, etc.). Nickel is also used in the production of non-ferrous alloys used, for example, in the manufacture of coins, tools and kitchen utensils.

There are three types of nickel oxide: nickel monoxide NiO, nickel dioxide NiO2 and nickel trioxide Ni2O3.

Nickel trioxide is a grey to black powder that can dissolve when heated in hydrochloric acid, releasing chlorine. Nickel trioxide and nickel dioxide are considered strong oxidants.

Nickel oxide NiO is obtained by heating nickel to a temperature in excess of 400° C. In the presence of oxygen, nickel is used in the production of special steels. This oxide is the only material used in dye-sensitized solar cells. Its electrical properties are as follows: its conductivity is of the order of 10^{-1}-10^{-3} s.cm^{-1}, its carrier density is of the order of 10^{17} cm^{-3} and its carrier mobility is 0.52 cm^2/s [48]. It is the only material available in the form of nanostructures in powder form. NiO exists in two allotropic forms, green and black, with the black form (Fig. 3-2) being the more reactive. It can react violently with iodine and hydrogen sulphide. Secondary ion mass spectrograms recorded on Fe, Ni and Cr show that the surface of these metals, when saturated with oxygen and bombarded with Ar^+ ions, is covered with an oxide film of the same composition as the bulk oxide FeO, NiO and Cr2O3 [49].

Figure 3-1: Nickel **Figure 3-2**: Nickel oxide

II. Experimental conditions

The experimental conditions will obviously be recalled whenever necessary. They can be described as follows:

> The kinetic energy of the bombarding ion beam is 5 keV, giving a travel speed for krypton of around 3.4 km/s. The use of krypton not only limits solid-gas reactions but also gives a higher sputtering rate than with argon, helium or neon ions [40]. In addition, krypton is less expensive than xenon and has fewer oxygen impurities.

> The intensity of the incident Kr beam+ is between 0.75 and 1.50 μA (i.e. approximately 10^{12} ions/ sec) with a cross-section of the order of 1 mm^2.

> The target is bombarded at an angle of 65 , the angle that the direction of the incident beam makes with the normal to the surface. This is the optimum angle for the best photon yield, as shown in Figure 1-5 in Chapter 1.

> The residual pressure in the enclosure containing the samples is better than 5.10 8 torr before the action of the beam. It increases to around 8.10 ^{8} torr during the bombardment and then, after a few minutes, stabilises to reach a value of around 10 7 torr.

With regard to the detection of light emissions :

> The spectral range explored extends from 190.00 to 650.00 nm.

> The minimum spectral resolution is set at 0.2 nm. A measurement step of 0.1 nm is used (since at least three points are needed to construct a line).

> The photomultiplier and counting electronics are adjusted at the start of each recording to give the optimum signal-to-noise ratio. This adjustment is generally made on the resonance lines and/or on the most intense lines.

> Before each recording, the target is bombarded for *in-situ* cleaning. The optical signal stabilises after a few minutes (5 to 10 min).

III. Experimental results

III.1. 1. Optical nickel emissions

The luminescence spectrum observed when a nickel sample was bombarded with Kr ions$^+$ of 5 keV is shown in Figure 3-3. It is recorded between 280 and 400 nm because outside this region, we did not observe any new lines apart from the second-order lines. The spectrum plotted in Figure 3-3 essentially comprises around twenty lines, all of which correspond to the de-excitation of neutral nickel atoms, i.e. Ni I. The most intense line is at wavelength λ=352.55 nm. This emission line, well known for this element, corresponds to the transition between excited states $3d^9 4p\ ^3P^\circ\ 2$ - $3d^9 4s\ ^3D_3$. It should be noted that no lines associated with Ni ions were observed and, moreover, we did not detect any continuous radiation in the spectral range explored. The characteristics of the spectral lines observed are shown in Table 3-1, the columns of which can be described as follows:

Columns 1 and 2 give the wavelengths of the observed X_o lines and those of the lines known from the literature X_L [50a,b] respectively.

Column 3 gives the relative intensities of the different lines observed in relation to that of the most intense line at 352.55 nm.

Column 4 shows the identification of the spectral lines.

Finally, columns 5 and 6 give the Einstein coefficients and lifetimes (in ns) [51] of the excited states corresponding to the observed transitions, when these are known.

Table 3-1 shows the following:

The recorded spectrum contains twenty-three spectral lines, all of which are identified with neutral Ni I atoms.

- Four of these twenty-three lines are associated with transitions to the ground state: $3d^8 4s^2\ ^3F_4$. These lines are located at 323.39, 337.05, 339.20 and 341.05 nm and have relative intensities of 9, 17, 25 and 60% respectively.
- Three lines are located at 341.05, 346.26 and 351.50 nm with relative intensities greater than 50% of that of the most intense line. The excited states of these lines have lifetimes of 15, 30 and 12 ns respectively.
- Twelve other lines have relative intensities of between 10 and 35% compared with the main line. These lines are located at 300.45, 305.17, 310.28, 337.05, 339.20, 343.45, 344.72, 345.39, 347.35, 349.47, 356.73 and 362.10 nm. These correspond to transitions between excited states.
- Seven lines have relative intensities of less than 10% compared with the most intense line. These lines are located at 313.50, 323.39, 342.47, 348.48, 359.88, 361.15 and 385.93 nm.

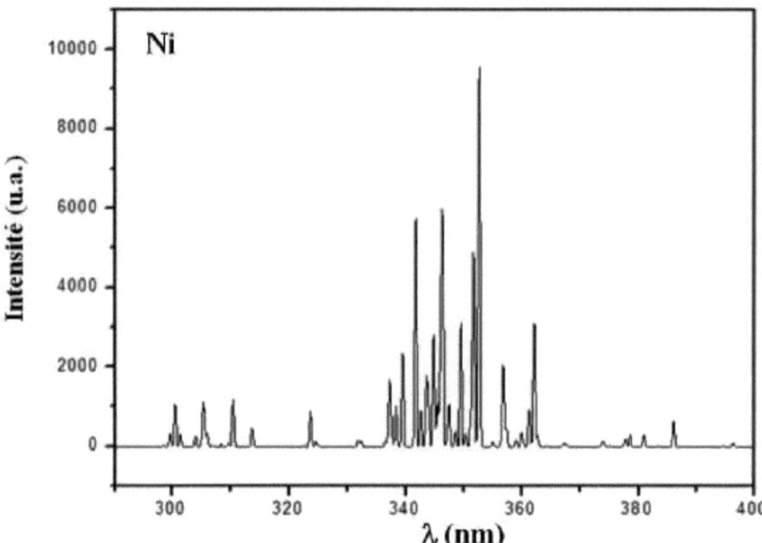

Figure 3-3: Optical emission spectrum of nickel

λo observed	/.' Literature[50a,b]	Ir(%)	Transition [50a,b]	Aki (10^8 s) [51]	τ (ns) [51]
300.50	300.45 Ni I	11	3d84s4p ^{3}D3 - 3d^{9}4s^{3}D2	0.69	-
305.30	305.17 Ni I	12	8 18 ∂ 3 3T7° 9 J 9 л 3y\ 3d 4s4p F 4- 3d 4s D3	0.6	-
310.40	310.28 Ni I	12	3d84s4p ^{1}F°3- 3d^{9}4s	0.49	-
313.60	313.50 Ni I	05	3d^{9}4p 1P°1- 3d^{9}4s ^{3}D2	0.73	-
323.50	323.39 Ni I	09	3d^{9}4p ^{1}P 1- 3d84s^2 ^{3}F$_4$	0.07	-
337.30	337.05 Ni I	17	3d^{9}4p^{3}G 4- 3d84s^2 ^{3}F4	0.02	17
339.40	339.20 Ni I	25	. .a. <20 _ .0 . n <2 3d^{9}4p ^{3}F 4- 3d84s^2 ^{3}F4	0.06	15
341.20	341.05 Ni I	60	A το . ϱO <2 3d^{9}4p ^{3}F 3- 3d84s^2 ^{3}F4	0.55	15
342.60	342.47 Ni I	09	3d^{9}4s^{3}D$_1$-3d^{9}4p^{3}D$_1$	0.33	-
343.60	343.45 Ni I	19	3d^{9}4p ^{3}F°3- 3d^{9}4s ^{3}D3	0.17	12
344.80	344.72 Ni I	29	3d^{9}4p ^{3}D°2- 3d^{9}4s ^{3}D2	0.44	17
345.50	345.39 Ni I	12	3d84s4p ^{5}F $_3$- 3d^{9}4s^{3}D$_2$	0.09	-
346.30	346.26 Ni I	63	3d84s4p ^{5}F°4- 3d4s ^{3}D3	0.27	30
347.50	347.35 Ni I	12	3d^{9}4p 1F°3- 3d^{9}4s ^{3}D2	0.12	11
348.60	348.48 Ni I	04	3d^{9}4p ^{3}D 1- 3d84s^2 ^{3}F2	0.14	-
349.50	349.47 Ni I	32	3d^{9}4p ^{3}P°1- 3d^{9}4s ^{3}D$_2$	0.98	10
351.70	351.50 Ni I	51	3d^{9}4p ^{3}F°3- 3d^{9}4s ^{3}D2	0.11	12
352.60	352.55 Ni I	100	3d^{9}4p ^{3}P°2- 3d^{9}4s ^{3}D3	1	-
356.80	356.73 Ni I	21	3d^{9}4p 1D°2- 3d^{9}4s 1D2	0.56	7.4
360.00	359.88 Ni I	04	3d^{9}4p ^{3}P°1- 3d^{9}4s ^{3}D1	0.14	10
361.30	361.15 Ni I	09	3d^{9}4p ^{3}P°2- 3d^{9}4s ^{3}D2	0.07	-
362.20	362.10 Ni I	32	3d84s^2 ^{3}F2- 3d84s4p ^{5}F 3	0.04	-
386.00	385.93 Ni I	06	3d^{9}4p ^{3}F°3- 3d^{9}4s 1D2	0.06	12

Table 3-1: Emission rays observed during bombardment of a sample clean nickel

III.2. 2. Optical emissions from oxygenated nickel

Figure 3-4a shows the luminescence spectrum that we recorded during the bombardment of nickel in the presence a pressure of 10 5 torr of oxygen. This spectrum is measured under the same bombardment and detection conditions as that of nickel at a base pressure of less than 10 7' torr. Comparison of the two spectra in Figures 3-3 and 3-4a, without and with oxygen, shows that we have the same number of lines, all at the same wavelengths, but with very different intensities. In fact, in the presence oxygen, we record an increase in all the spectral lines associated with Ni I. The most intense line in the spectrum of oxygenated nickel (Fig. 3-4 a) only increased by a factor of 1.18. The lines associated with transitions to the ground state ($^3F_4$) underwent an increase of 1.53, 1.33, 1.35 and 1.31, which we have noted by the ratio R_1 in Table 3.2. These lines are located at 323.39, 337.05, 339.20 and 341.05 nm respectively.

It should be noted that in the spectral region explored, we did not observe any emission attributed to oxygen. In addition, as with the clean Ni spectrum, we did not record any lines associated with Kr^+ ions or continuous radiation.

III.3. 3. Optical emissions of nickel oxide

Figure 3-4b shows the luminescence spectrum recorded between 280 and 400 nm of nickel oxide (NiO) target during bombardment by Kr ions$^+$ of 5 keV. The lines obtained in this case are exactly the same as those measured for clean nickel. However, the intensities decreased by a factor of about 4 for most of the spectral lines. The most intense line is still located at 352.55 nm, which we identify with the ^{3}P transition$^\circ$ 2-$^3D_3$.

The R_2 ratio of spectral line intensities (Table 3-2) reveals the following points:

- Two lines, located at 347.35 and 348.48 nm, underwent a significant decrease, with R_2 reaching a factor of 0.02.
- Four lines, located at 323.39, 337.29, 339.20 and 341.05 nm, associated with transitions to the ground state ($^3F_4$) decreased by almost the same R2 ratio, close to 0.30.
- The line at 361.15 nm has an intensity that seems to remain insensitive, since it has an R2 ratio equal to 1.09.
- Three lines have intensities that decrease by a ratio R_2 of the order of 0.75; they are located at 300.45, 362.10 and 385.93 nm.
- Three lines have intensities with an R2 ratio that varies between 0.20 and 0.30. These lines are located at 305.17, 310.28, 313.50, 342.47, 343.45, 344.72, 345.39, 346.26, 349.47, 351.50, 352.55, 356.73 and 359.88 nm.

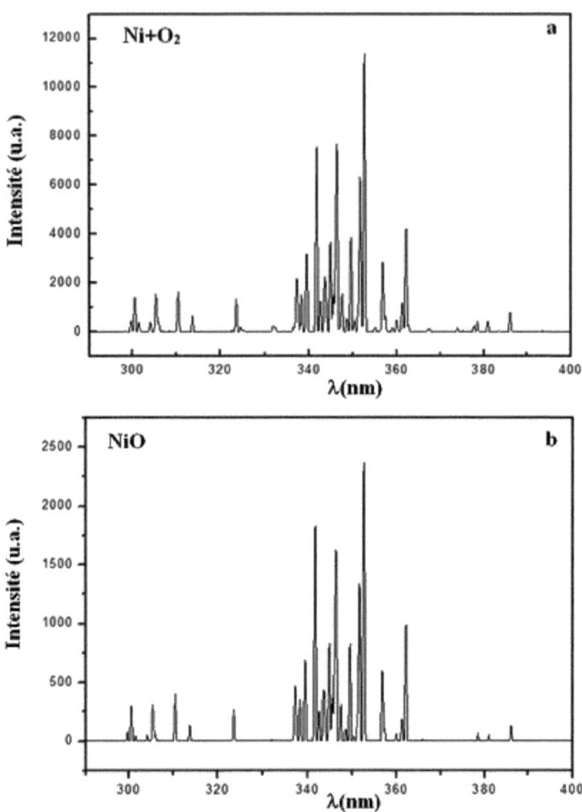

Figure 3-4: Luminescence spectra recorded during ion bombardment of samples of: a) nickel in the presence of oxygen, b) nickel oxide.

Ni ([Ar] $3d^8$ 4s2- ^{3}F4)

λ (nm) Literature	Absolute current I (a.u.)			Intensity ratios		Level energy (eV)		
	Ni	Ni+O2	NiO	$R_1 = \frac{I_{Ni+O_2}}{I_{Ni}}$	$R_2 = \frac{I_{NiO}}{I_{Ni}}$	ESup	Ejnf	Ev
300.45 Ni I	1060	1405	800	1,32	0,75	4.15	0.10	-3.48
305.17 Ni I	1120	1555	315	1,38	0,28	4.08	0.02	-3.55
310.28 Ni I	1045	1565	405	1,49	0,38	4.42	0.42	-3.21
313.50 Ni I	455	655	145	1,43	0,31	3.83	0.21	-3.80
323.39 Ni I	885	1360	270	1,53	0,30	3.83	0.00	-3.80
337.05 Ni I	1630	2175	470	1,33	0,28	3.67	0.00	-3.96
339.20 Ni I	2355	3185	700	1,35	0,29	3.65	0.00	-3.98
341.05 Ni I	5740	7570	1840	1,31	0,32	3.63	0.00	-4.00
342.47 Ni I	900	1245	260	1,38	0,28	3.83	0.21	-3.80
343.45 Ni I	1785	2250	430	1,26	0,24	3.63	0.02	-4.01
344.72 Ni I	2810	3695	835	1,31	0,29	3.70	0.10	-3.93

345.39 Ni I	1120	1505	366	1,34	0,32	3.69	0.10	-3.94
346.26 Ni I	5990	7695	1630	1,28	0,27	3.60	0.02	-4.03
347.35 Ni I	1065	1555	30	1,46	0,02	3.68	0.10	-3.95
348.48 Ni I	355	535	10	1,50	0,02	3.83	0.27	-3.80
349.47 Ni I	3100	3885	840	1,25	0,27	3.66	0.10	-3.97
351.50 Ni I	4905	6340	1345	1,29	0,27	3.63	0.10	-4.00
352.55 Ni I	9565	11380	2375	1,18	0,24	3.54	0.02	-4.09
356.73 Ni I	2045	2875	605	1,40	0,29	3.90	0.42	-3.73
359.88 Ni I	360	485	65	1,34	0,18	3.66	0.21	-3.97
361.15 Ni I	910	1185	992	1,30	1,09	3.54	0.10	-4.09
362.10 Ni I	3080	4228	2330	1,37	0,75	3.85	0.42	-3.78
385.93 Ni I	620	810	465	1,30	0,75	3.63	0.42	-4.00

Table 3-2: Emission rays observed during bombardment of the samples Ni, Ni+O2 and NiO

IV. Discussion and interpretation of results

In Table 3-2, we have reported the characteristics of the spectral lines whose absolute intensities are greater than 3% of that of the most intense line (λ= 352.60 nm). The first column gives the wavelengths of these lines, which we have identified as Ni I atomic lines. The next 3 columns give the absolute intensities of these lines measured when bombarding nickel, nickel in the presence of oxygen (oxygenated nickel) and nickel oxide NiO. In the 5th and ^{6th} columns, we have indicated the intensity ratios between oxygenated nickel and clean nickel (R_1 ratio) and between nickel oxide and clean nickel (R_2 ratio) respectively. The last three columns give the energy positions of the upper and lower states (in eV) as well as the energy E_v defined by E^*-I_+ where E^* is the energy of the emitting state and I_+ is the simple ionisation potential $I_+ = 7.63$ eV **[**50a,b**].** In the presence of oxygen, we noted an increase in the intensities of the spectral lines and we say that these lines have a positive dependence. On the other hand, we recorded a decrease in the intensities of these same lines in the case of nickel oxide. Moreover, we must point out that the intensity of the bombardment ion beam was 0.9 µA for NiO whereas it is 1.16 µA for bulk Ni. This difference in spectral intensity was also observed in the case of bulk Fe and powdered magnetite [11]. It is noted that in order to establish a comparison between the spectra, it would be useful to carry out the experiments with samples of the same nature, namely Ni and NiO both in powder form.

The exaltation of intensities observed for nickel in the presence of oxygen (oxygenated nickel) has also been reported by Kaddouri [40] and van der Weg et al [52]. This exaltation can be adequately explained by the electron exchange model. To illustrate this model, we have plotted an energy level diagram in Figure 3-5 using vacuum energy as the origin. The left-hand column shows the energy bands of the metal Ni and the right-hand column those of its oxide NiO. For information purposes, we have plotted between these bands some of the energy levels of the excited atom that are associated with certain optical transitions observed. The clean metal is thus identified by its work function φ(Ni) =5.15 [53] and the NiO oxide by its electron affinity A_f of value 1.46 eV [54] and by its gap E_g 3.3eV [55] and finally the state of the excited atom is identified by the energy E_v. In the electron exchange model, the oxygen adsorbed on the surface of the target modifies the energy band structure the initial target and the solid thus formed is an "oxide" characterised by a gap that separates the conduction band from the valence band. Referring to the energy diagram in Figure 3-5, we note that the emitting states of the lines, manifesting a positive dependence, are all located opposite both the conduction band ($-\Phi<E_v<0$) and the oxide gap ($E_v<-A_f$). Such an energy position entails that these excited states can then exchange their excited electron with the metal and not with the oxide formed. Consequently, the presence a gap blocks the non-radiative de-excitation processes and therefore allows the excited states of the particles to be depopulated by photon emission, which, in the case of oxygenated nickel, intensifies the light signal from the lines associated with these emitting states. As mentioned above, all the line intensities underwent an increase (positive dependence), which shows that all the emitting states in these lines must be opposite the gap in the oxygenated Ni sample. This model suggests that the presence oxygen in the vicinity of the clean metal leads to the formation a layer made up metal atoms and oxygen in which the metal-oxygen chemical bond is fairly close to that of the oxide. It should be noted, however, that the composition of this new layer is not necessarily that of the bulk oxide, and can vary with depth.

As for the nickel oxide sample, which we used in powder form, we measured a decrease in spectral line intensities. This decrease is probably linked to the amount of material sputtered, since the latter is governed by the efficiency of the cascade collisions initiated by the incident ion. The energy transferred during the collisions appears to be 'annihilated' in the powdered nickel oxide, leading to a reduction in the number of particles ejected and therefore in the number of species excited. This result is in agreement with the work of Hammoum et al [11] on solid iron and magnetite (Fe_2O_3) in powder form.

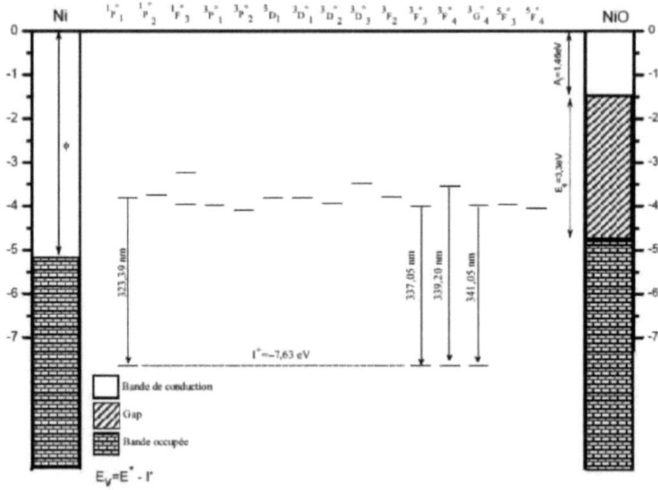

Figure 3-5: Diagram of the energy levels of the Ni metal, the NiO oxide and the emitting states of the Ni I lines

Part 2

Optical emissions from chromium and its oxide Cr2O3

I. Physical and chemical properties of chromium

Chromium belongs to the transition metal series, its atomic number is Z= 24 and its electronic configuration is given by : [Ar] $3d^5$ $4s^1$ corresponding to the spectral term $^7s_3$. Chromium occurs naturally as a hard, steel-grey metal (Fig. 3.6).

Chromium oxide Cr_2O_3 is a crystalline solid obtained by the action of sulphuric acid on potassium dichromate. It has the appearance of a green powder (Fig. 37). This colouration is due to the excitation of the energy levels of the electrons in the d sublayer by certain wavelengths of light. Cr_2O_3 is characterised by its good hardness and chemical stability at room temperature. It is very stable and occurs naturally in ores such as chromite. Cr_2O_3 is also widely used as a protective and electrochemical coating. However, unlike Cr_2O_3, chromium oxide CrO_3 is highly unstable and oxidises rapidly to Cr_2O_3. The kinetics and toxicity of chromium compounds depend mainly on the oxidation state of the compound and its solubility in biological media. The main users of chromium and its oxides are metallurgy (stainless steel, ferrous and non-ferrous alloys), the refractory sector (resistance to high temperatures) and the chemical industry (pigments, chromium plating, tanning, wood preservation).

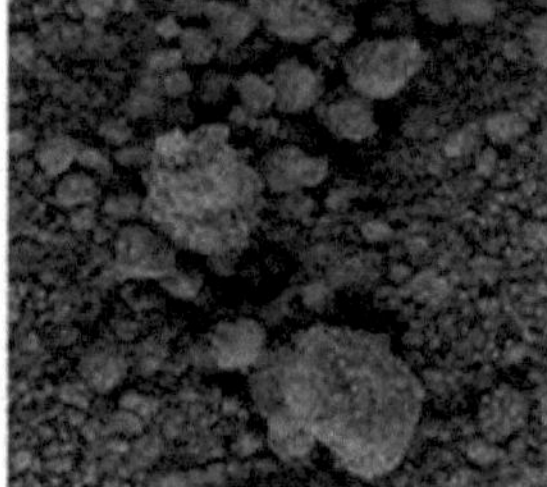

Figure 3-6: Solid chromium **Figure 3-7**: Chromium oxide

In 1797, chromium oxide (Cr_2O_3) was produced by adding hydrochloric acid to chromite. A year later, it was possible to isolate metallic chromium by heating the oxide in a coal-fired furnace. In 1953, Shull and Wilkmson [56a] demonstrated the antiferromagnetism of chromium by neutron diffraction. They showed that the metal has a magnetic structure in which the atoms at the vertices and those at the centre of the elementary cubic lattice have antiparallel magnetic moments. Its melting temperature is 1907° C and its boiling temperature is 2671° C and its density is 7.15 g cm '3 at 20 C [56b].

Cr2O3 oxide is characterised by a gap whose value is widely disputed. Measurements carried out on a passive chromium oxide film using UV photoluminescence excitation show a peak at 400 nm for an acid solution and at 440 nm for a neutral solution. These values correspond to gap values of 3.1 eV and 2.8 eV respectively [57]. Misho et al [58] also studied the optical properties of a Cr2O3 oxide film obtained by the spray pyrolysis method by spraying a $Cr(NO_3)_3$ solution onto a glass substrate preheated to a temperature of 500° C. Analysis of the absorption and transmission spectra of the film gave a gap value of 3.40 eV [58]. The study of the effect of temperature on the optical properties was recently repeated by Al-Kuhaili and Durrani [59] on a chromium oxide film deposited by electron beam evaporation. Their work revealed a gap of 2.86 eV for a film deposited on a substrate heated to 300° C and a gap of 2.67 eV when deposited on an unheated substrate. Hones et al [60] measured gap values ranging from 3.62 eV to 3.71 eV for oxide films deposited by sputtering at different temperatures and different oxygen pressures. In the case of a pollycrystalline Cr_2O3 film deposited by chemical vapour deposition, Julkarnain et al [61] determined a gap of 3.09 eV at T=720 K and 2.98 eV at T=820 K.

Concerning the spectroscopic study, the optical emission of chromium was analysed using glow discharge spectroscopy by Wagatsuma [62], where 2049 lines were observed in a spectral range extending from 200 to 440 nm. In addition, work by MacDonald and Martin [63] has shown that ion bombardment of chromium in an oxygen atmosphere improves the yield of monocharged Cr^+ ions from the polycrystalline target, using optical emission and emission of secondary chromium ions.

We will see that our spectra contain very few spectral lines compared with the spectra obtained by the glow discharge, due to the fact that under our experimental conditions, the plasma produced by the ion beam is not hot.

II. Experimental results

II.1. . Optical emissions of chromium

The experimental bombardment conditions are the same as before (Part A). With regard to detection, the spectral range explored still extends from 190.00 to 650.00 nm, but the vast majority of lines are located between 350 and 470 nm.

The luminescence spectrum of chromium obtained during bombardment by Kr $ions^+$ of 5 keV is shown in Figure 3-8. It contains around fifteen relatively intense discrete lines. The spectral lines observed are directly identified with Cr I lines, i.e. optical emissions from neutral atoms. It should be noted that no lines associated with Cr ions are observed and, moreover, we did not detect any continuous radiation in the spectral range explored.

In the recorded spectrum, the most intense line is located at 425.50 nm and is associated with the transition of a 4p electron to the 4s orbital of the ground state, i.e.: $3d^5 4p\ z^7P^\circ\ 4$ - $3d^5 4s\ a^7s_3$. The excited state lifetime is 31.6 ns. For evaluation purposes, we have limited ourselves to examining only those lines whose intensity is greater than 3% of that of the most intense line.

The wavelengths of the lines detected are given in Table 3-3, together with their relative intensities and the identification of the optical transitions involved. The last columns give the Einstein coefficients and lifetimes of certain excited states.

In addition to the most intense line, we have five other relatively intense lines located at 357.97, 359.45, 360.60, 427.60 and 429.09 nm with relative intensities of 48, 36, 26, 93 and 67 % respectively. All these lines correspond to transitions to the ground state: $^7s_3$. The emitting states of these lines have lifetimes of 6.9, 7.1, 6.3, 32 and 32 ns respectively.

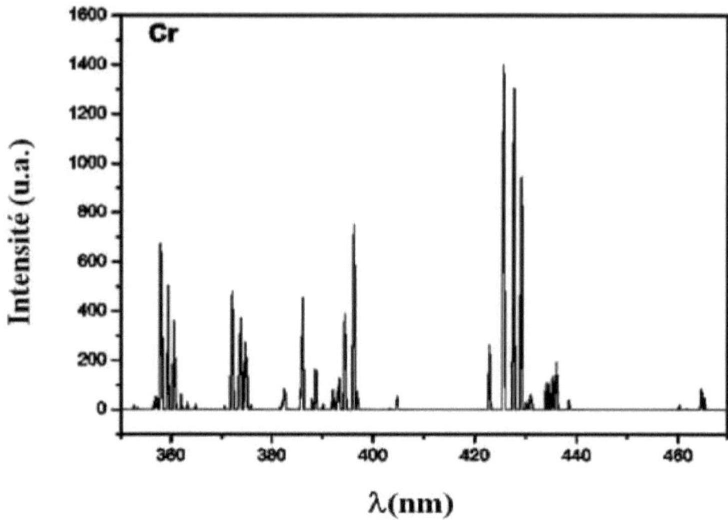

Figure 3-8: Optical emission spectrum of chromium
Cr ([Ar] $3d^5 4s^1$ - $^7S_3$)

λ_O Observed	λ_L Literature[50a,b]	Ir(%)	Transition [50a,b]	Aki (108 sec) [51]	τ (ns) [51]
357.90	357.97 Cr I	48	3d44s4py $^7P°_4$ - $3d^54sa\ ^7S_3$	0.69	6.9
359.40	359.45 Cr I	36	3d44s4py $^7P°_3$ - $3d^54sa\ ^7S_3$	0.6	7.1
360.60	360.63 Cr I	26	3d44s4py $^7P°_2$ - $3d^54sa\ ^7S_3$	0.49	6.3
388.70	388.63 Cr I	12	3d44s4pz $^5D°_2$- $3d^44s^2a\ ^5D1$	0.73	-
392.00	392.03 Cr I	04	3d44s4pz $^5D°_4$ - $3d^44s^2a\ ^5D_4$	0.02	-
393.00	392.98 Cr I	05	3d44s4pz $^5D°_2$ - $3d^44s^2a$ $5D_3$	0.55	-
404.70	404.33 Cr I	04	3d44s4px $5F°_3$ - $3d^54sa\ ^5G_4$	0.33	-
425.50	425.55 Cr I	100	$3d^54p$ z $^7P°_4$- $3d^54s$ a^7S_3	0.17	31.6
427.60	427.60 Cr I	93	$3d^54pz\ ^7P°_3$ - $3d^54s$ a $^7S_3$	0.44	32
429.10	429.09 Cr I	67	$3d^54pz\ ^7P°_2$ - $3d^54sa\ ^7S_3$	0.09	32
434.60	434.43 Cr I	08	3d44s4pv $5P°_2$ - $3d^54sa\ ^5P1$	0.27	-
435.30	435.29 Cr I	10	3d44s4pz $^5F°_5$ - $3d^44s^2a\ ^5D_4$	0.12	-
438.40	438.41 Cr I	03	3d44s4pv 5P°1 - $3d^54sa$ T'-	0.14	-
464.70	464.74 Cr I	06	3d44s4py $^5P°_3$ - $3d^44s^2a\ ^5D_4$	0.98	63
465.30	464.96 Cr I	04	$3d^54py\ ^5H_6$- $3d^44s^2a\ ^3H_6$	0.11	9.5

Table 3-3: Chromium optical emission spectra and their identification

II.2. . Optical emissions from oxygenated chromium and chromium oxide

The luminescence spectra observed in the 350 to 470 nm region when chromium samples were bombarded in an oxygen atmosphere of 10 5 torr and chromium oxide are shown in figures 3-9 a and b. They show a series of discrete lines at the same wavelengths as those obtained with a clean chromium sample and are therefore all identified as Cr I lines. In the two spectra recorded, no lines associated with an ion are observed in the spectral range explored. Similarly, no optical emission of oxygen or krypton was detected. Also, in the spectral region explored, we did not observe any continuous radiation, in contrast to a study by Kiyan et al [64] who found continuous optical emission when chromium was bombarded by Ne^+ , Ar^+ and Xe^+ ions with a kinetic energy of 30 keV. The two spectra differ slightly in intensity from that of clean chromium. We also noted the presence of four spectral lines at

362.00, 372.10, 374.70 and 382.60 nm, which we identified as iron lines. The detection of these foreign lines shows that the SIPS technique we use is very sensitive to the presence of trace impurities. The intensity of these lines decreases as the sample is bombarded. These lines eventually disappear and will not be reported in the tables or discussed below. Comparison of the recorded spectra shows that the intensities of the Cr I spectral lines are enhanced both for chromium in the presence oxygen (oxygenated chromium) and chromium oxide Cr_2O_3. However, this enhancement is much more pronounced in the case of oxygenated chromium. This increase depends on the wavelength and therefore on the electronic transition involved; for example, we noted an increase in the intensities of the lines measured at 357.97 and 435.29 nm with a factor of 1.86 and 2.92 respectively. The same observation was made by Kaddouri [40] in the case of solid chromium, who experimental conditions very similar to ours.

Table 3-4 gives the wavelengths of the Cr I lines (column 1) and the absolute intensities (column 2) of these lines observed during the bombardment of clean chromium, oxygenated chromium and chromium oxide. Column 3 gives the ratio of the absolute intensities. The last column gives the energy positions (in eV) of the upper and lower electronic states involved the transition under consideration, as well as the E_v energy that we will use in the discussion in the next paragraph. This energy is defined relation to vacuum energy and is equal to the difference between the energy E^* of the emitter state and the ionisation potential I^+ of chromium, which is 6.76 eV [50a,b].

In Table 3-4, the column giving the R_1 ratio reveals the following points:

- The intensities of the spectral lines observed in the case of oxygenated chromium underwent a significant increase, with the R_1 ratio varying between 1.41 and 5.75.
- The main line, located at 425.55 nm, has increased in intensity by a factor of 1.57.
- Six atomic lines have an R_1 ratio varying from 1.40 to 1.95. These lines de-excite towards the ground state ($^7S_3$) and are located at 357.97, 359.45, 360.63, 425.55, 427.60 and 429.09 nm.

As for the column giving the ratio R_2 associated with the oxide, we can note that :

- The main line only increased by a factor of 1.12.
- The majority of lines increased by an R_2 ratio varying between 1.05 and 1.58.
- The Cr_2O3 spectral line located at 427.60 nm has an intensity that seems to remain insensitive since it has an R_2 ratio equal to 1.03.

The intensity of all the lines emitted by Cr2O3 decreased compared with those emitted by oxygenated chromium, and this decrease reached 60%.

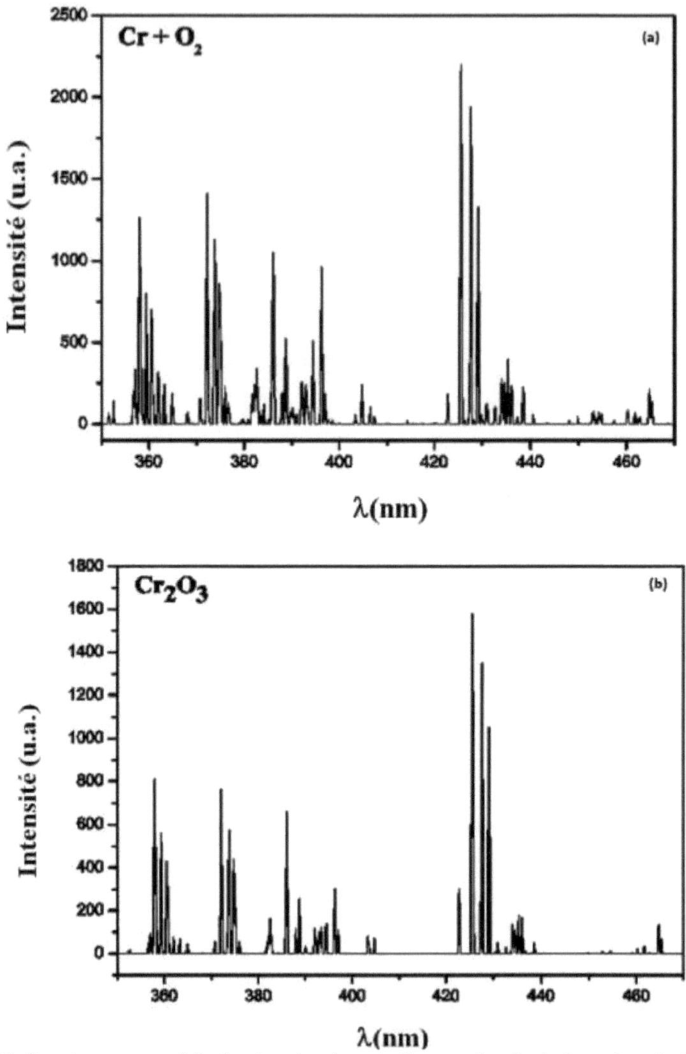

Figure 3-9: Spectra measured during bombardment of a sample of: a) chromium in the presence of oxygen, b) chromium oxide

Cr ([Ar] $3d^4 4s^2$- $^7S_3$)

λ (nm) Literature	Absolute current (a.u.)			Intensity ratios		Level energy (eV)		
	Cr P=10"7 (torr)	Cr+O2 P=10"5 (torr)	Cr2O3 P=10"7 (torr)	$R_1 = \frac{I_{Cr+O_2}}{I_{Cr}}$	$R_2 = \frac{I_{Cr_2O_3}}{I_{Cr}}$	E ^sup	Einf	Ev
357.97 Cr I	675	1260	810	1,86	1,20	3.463	0.000	-3.30
359.45 Cr I	500	805	560	1,61	1,12	3.449	0.000	-3.32
360.63 Cr I	360	705	430	1,95	1,19	3.438	0.000	-3.33
388.63 Cr I	165	520	255	3,15	1,54	4.159	0.968	-2.61
392.03 Cr I	80	260	120	3,25	1,50	4.193	1.030	-2.57
392.98 Cr I	75	240	100	3,20	1,33	4.159	1.004	-2.61
404.33 Cr I	55	240	70	4,36	1,27	5.611	2.545	-1.16
425.55 Cr I	1400	2200	1580	1,57	1,12	2.913	0.000	-3.85
427.60 Cr I	1300	1940	1345	1,49	1,03	2.899	0.000	-3.87
429.09 Cr I	940	1330	1050	1,41	1,11	2.889	0.000	-3.88
434.43 Cr I	105	250	110	2,38	1,05	5.564	2.710	-1.20
435.29 Cr I	135	395	180	2,92	1,33	3.878	1.030	-2.89
438.41 Cr I	40	230	55	5,75	1,37	5.538	2.710	-1.23
464.74 Cr I	85	215	135	2,52	1,58	3.698	1.030	-3.07
464.96 Cr I	50	140	70	2,80	1,40	5.667	3.000	-1.10

Table 3-4: Emission rays observed during bombardment of a sample of chromium in the presence and absence of oxygen

III. Discussion and interpretation of results

We have recorded an increase in the intensities, i.e. the positive dependence, of the Cr I lines when moving from clean Cr to oxygenated Cr. In an attempt explain this behaviour, Figure 3-10 shows a schematic diagram of the energy levels of the metal Cr, the oxide Cr_2O_3 and the higher levels associated with some of the observed optical transitions. In this diagram, the metal is characterised by its work function φ = 4.5 eV [65]; the chromium oxide Cr_2O_3, being a semiconductor, is characterised by the width its band gap: E_g = 2.67 eV [59] and by the electronic affinity that we will estimate later. Finally, the emitting states of excited atoms are identified by the energy E_v.

According to the model, the lines that manifest the positive dependence must have their emitter state located opposite both the metal's work function band (0< py|< Φy) and the oxide's gap (Af <|EV|<Af + Eg). These states can then exchange their excited electron with the metal and not with the oxide. As a result, the non-radiative de-excitation processes are blocked and the depopulation of the excited states then takes place by photon emission, which intensifies the light signal from the lines associated with these emitting states.

These considerations allow us to suggest, for oxygenated chromium, a gap value of at least 2.78 eV (E_g is determined by the energy difference between the $|E_v|$ value corresponding to the 3.88 eV energy level associated with the 429.09 nm line and the lowest E_v value of 1.10 eV associated with the 464.96 nm line). The value of the band gap that we measured is in perfect agreement with the value of E_g=2.67 eV obtained by Al-Kuhaili and Durrani [59] for a chromium oxide film deposited on an unheated substrate. As for the electronic affinity, we can suggest a maximum value of 1.10 eV. This value is the smallest of the absolute $|E_v|$

energies and is associated with the 464.96 nm line showing a positive dependence.
These results lead us to propose the energy band diagram for oxygenated Cr which we have plotted on the same figure 3-10. This diagram assumes that the presence of oxygen "drastically" modifies the energy band structure of the initial target following chemisorption of this gas. This assumption can be supported by the work of Sakisaka et al [66] who studied the interaction of oxygen at 300 K with the chromium surface using AES (Auger spectroscopy), LEED (slow electron diffraction) and SES (secondary electron spectroscopy) techniques. They showed that the variation in work function as a function of oxygen pressure reveals the formation of a Cr_2O_3 oxide layer.
In our experiments, the new solid formed is an oxide characterised by a band gap and electron affinity. The diagram we have proposed for the oxide film formed adequately explains the observed exaltation of all the Cr I spectral lines when switching from the clean metal to the oxygenated metal. In fact, we can consider that under our experimental conditions the behaviour observed for light emission in the presence of oxygen is governed by a situation in which a system with a composition and band structure intermediate between that of the metal and that of the oxide is formed. This system is a "metal-oxygen" mixture whose strechiometry is different from that of the oxide and is governed by the competition between the erosive action of the bombarding ion beam and the adsorption of oxygen atoms onto the surface.
The decrease in line intensities that we observed when moving from oxide to oxygenated chromium is probably due to the contribution of a mechanism that is more favourable in the case of oxygenated Cr than for oxide. This is the radiative dissociation of $Cr\text{-}O_2$ molecules. These molecules are formed by adsorption of oxygen on the Cr surface and are ejected, under the effect of bombardment, at a distance from the surface to escape the non-radiative process due to interaction with the surface.

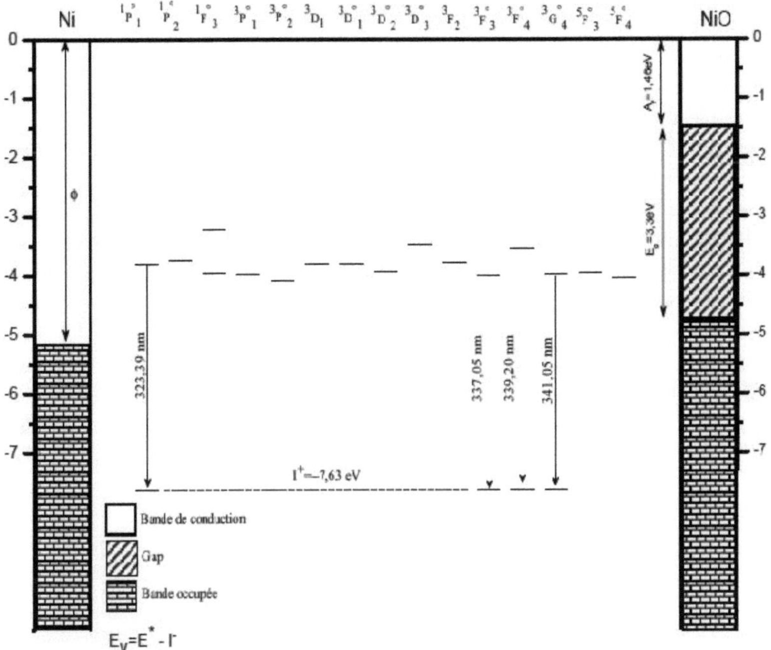

Figure 3-5: Diagram of the energy levels of the Ni metal, the NiO oxide and the emitting states of the Ni I lines

Conclusion

We studied the luminescence spectra of the following two transition metals: nickel and chromium when bombarded with Kr^+ ions at 5 keV. We also measured the luminescence spectra of the corresponding oxides under the same experimental conditions: NiO and Cr_2O_3. The optical radiation observed is in the form of discrete lines attributed to electronic transitions characteristic of neutral nickel and chromium atoms, i.e. Ni I and Cr I lines. Our results also show that the presence of oxygen during target bombardment gives the same spectral lines but with an increase in intensity. These same spectral lines are also present case of an oxide target with intensities higher than those obtained for clean metal but lower than those for oxygenated metal. For these two elements, we did not observe any continuous radiation.

The exaltation of the intensities of the spectral lines is interpreted appropriately within the framework of the electron exchange model between the surface and the ejected particle. The validity of this model suggests that, in the presence of oxygen, a structure is formed whose energy band pattern is intermediate between that of the metal and that of the oxide. The slight increase in spectral intensities observed in the case of oxygenated metal compared with oxide is explained by the contribution of radiative dissociation.

Chapter 4

Optical emissions from for spraying titanium

Introduction

The results presented in this chapter were obtained by bombarding a solid titanium target with Kr^+ ions with a kinetic energy of 5 keV.

We will examine the behaviour of the luminescence spectra of titanium in a vacuum of better than 10 7 torr and of titanium in the presence of a controlled atmosphere of oxygen at a pressure of 10 5 torr (oxygenated titanium).

The luminescence spectra we obtained consist of a series of very fine lines superimposed on a continuum. It should be noted that in the spectral region explored (190-650 nm), no emission lines attributed to oxygen were observed. All the spectral lines recorded were identified as Ti I neutral and Ti II ionic atoms.

In the presence of oxygen, the intensities of the spectral lines do not evolve in the same way. In fact, we observed that the intensities of the Ti I lines showed an increase and insensitivity, while the Ti II lines showed a decrease.

We will interpret our results in the context of the electron exchange model, which assumes the formation of an oxide layer very close to the bulk oxide. This oxide is defined by an electron affinity and a band gap.

Before presenting these results, we will review some of the physical and chemical properties of titanium and its oxide.

I. Physical and chemical properties of titanium

Titanium is a metal with a compact hexagonal crystallographic structure (a = 0.285 nm; c = 0.468 nm; c/a = 1.633) and a grey colour. Its atomic mass is 47.88. The electronic configuration of titanium is given by [Ar] $3d^2 4s^2$ with $^3F2$ as the ground state spectral term. Its ionisation potential is 6.83 eV.

Titanium is considered to be a metal with high mechanical strength and good ductility under normal temperature conditions, and is one of the most biocompatible metals along with gold and platinum. This metal has many applications: it is used in the composition of alloys used in surgery and in the manufacture of everyday consumer goods because it is hard, light, resistant and non-toxic [67]. It does not react with water many acids, but is more sensitive to hydrogen and oxygen, which absorb onto its surface.

Like many transition metals, titanium can be found in several oxidation states. It therefore has several oxides corresponding to different degrees of oxidation: titanium monoxide (TiO), dititanium monoxide (Ti_2O), trititanium monoxide (Ti_3O), titanium dioxide (TiO_2), titanium trioxide (TiO_3) and dititanium trioxide (Ti_2O_3).

Titanium dioxide TiO2 is the most widely used white pigment and exists in four phases (Fig. 4-1): rutile, anatase, brookite and β-TiO_2. Depending on the nature of the phase, the physico-chemical properties of titanium dioxide differ because their crystalline structure changes. The band gap energy of anatase is 3.23 eV and that of rutile is 3.02 eV [68]. The position of the valence band allows the formation of energetic holes and easy oxidation of adsorbed species. Buchalska et al [69] have shown that UV absorption properties are only effective with the anatase and rutile forms, given their absorption capacity for a wavelength of around 400 nm.

Parker and Kelly [70] have shown that bombardment of TiO_2 by krypton ions with a kinetic energy of 30 keV leads to the formation of a new oxide layer with a high electrical conductivity. The formation of this layer is enhanced by the increase in target temperature.

TiO_2 has a wide range of applications. Firstly, it is used in biomaterials, in particular for biomedical implant devices. Secondly, as an antibacterial agent. It is currently used as an important element in the study of metal hydrides using impedance spectroscopy [71] and also as a cleaner and/or disinfectant for certain surfaces due to its photocatalytic properties. In 2014, global consumption of titanium dioxide reached 5.5 million tonnes. Since 1980, when consumption was around 2.1 million tonnes, both demand and production have increased fairly linearly.

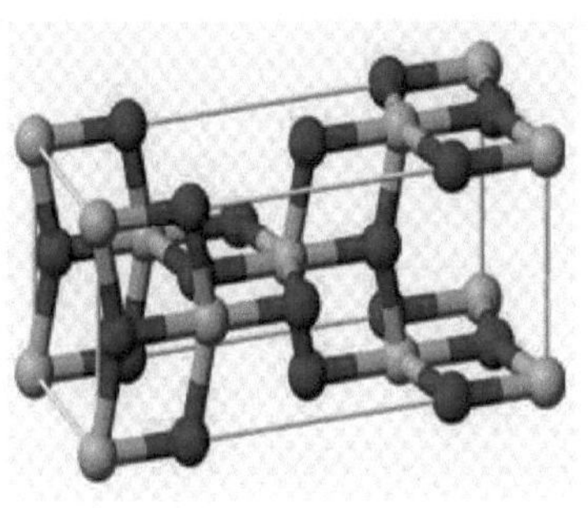

Structure cristallographique de l'anatase

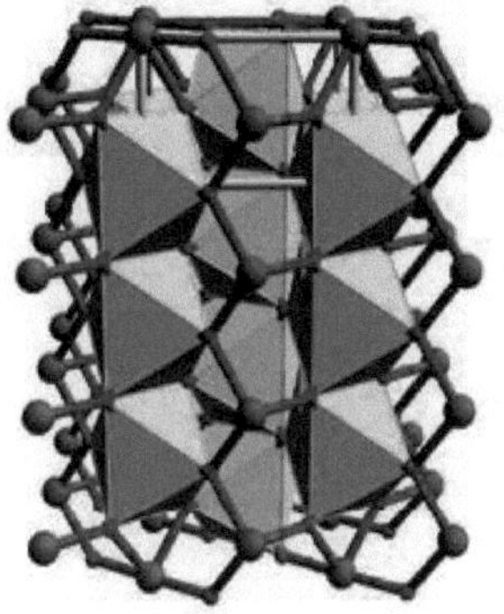

Structure cristallographique de brookite

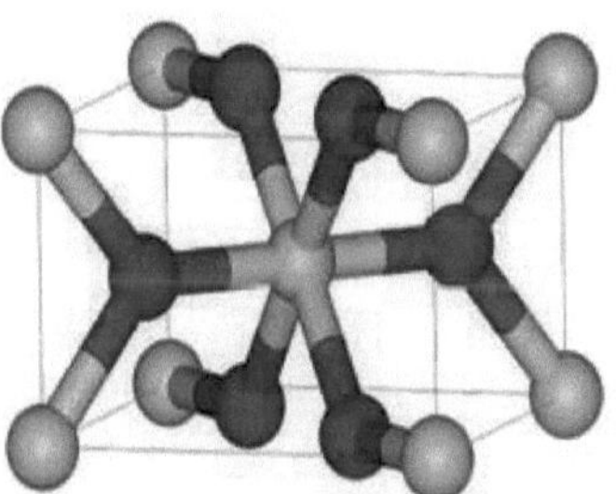

Structure cristallographique de rutile

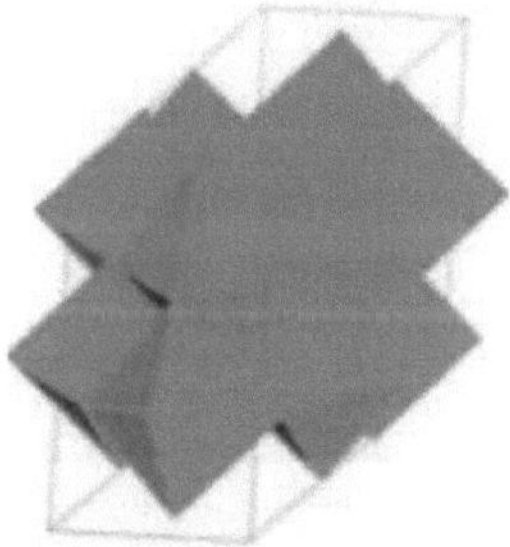

Structure cristallographique β-TiO_2

Figure 4-1: Titanium dioxide phases

II. Experimental conditions

The results we are about to present were obtained using the SIPS apparatus described in the second chapter. These results concern the optical emission spectra of titanium samples and titanium in the presence of oxygen. Each spectrum was obtained after several recordings, often several days apart, in order to check the reproducibility of the recorded spectra.

The experimental conditions used are :

> The kinetic energy of the bombarding ion beam is 5 keV, giving a travel speed for krypton of around 3.4 km/s.

> The target is bombarded at an angle 65 , the angle between the direction of the incident beam and the normal to the surface. This value corresponds to the maximum light signal

associated with the spectral lines observed.

> The incident beam from Kr^+ has an intensity of between 0.75 and 1.50 μA and a cross-section of the order of $1 mm^2$.

> The residual pressure in the enclosure containing the samples was measured a few cm from the sample and was better than 8.10 8 torr before spraying. After a few minutes of bombardment, the pressure stabilised to reach a value of around 10 7 torr.

> Bombardment in the presence of oxygen takes place at a pressure of 10 5 torr of this gas.

With regard to the detection of light emissions ;

> The spectral range of detection sensitivity extends from 190.00 to 650.00 nm.

> The minimum spectral resolution is set at 0.2 nm in order to have enough points to construct a spectral line.

> The optical axis of the monochromator is fixed perpendicular to the direction of the incident ion beam.

III. Presentation of results

III.1. 1. Optical emissions from clean titanium

Figure 4-2 shows the luminescence spectrum observed when clean titanium is bombarded by krypton ions with a kinetic energy of 5 keV. This spectrum is shown between 280 and 420 nm because outside this region we did not observe any new spectral lines. The wavelength values given are those read directly from the monochromators, and the reported intensities are normalised to the most intense radiation. We found around twenty discrete lines superimposed on a continuum extending from 275 to 400 nm and centred at 325 nm. This continuum will be described and interpreted in the last paragraph.

Identification of the lines observed revealed that they were optical emissions from neutral Ti I atoms and simply charged Ti II ions. No optical emission from oxygen or krypton was detected. The most intense line (known as the main line) is located at 374.27 nm and corresponds to the transition between excited states $3d^2 4p\ y^2 D_{5/2}$- $3d^3\ b^2 D^2{}_{5/2}$ and a Ti II line. The same observations were made by Varene [27] who used experimental conditions very similar to ours.

Table 4-1 shows the characteristics of the lines detected whose intensities are greater than 7% of that of the main line. The columns of this table can be described as follows:

Columns 1 and 2 give the wavelengths of the observed lines λ_σ and those of the lines known from the literature X_L [50a,b].

Column 3 gives the relative intensities of the different lines observed in relation to the most intense line.

In the last two columns, we have reported the identification of the lines observed and the corresponding Einstein coefficients, respectively.

This table gives the relative intensities of twenty-four spectral lines, thirteen of which are Ti I lines associated with neutral atoms and eleven Ti II lines attributed to simply charged ions. Of these twenty-four lines, six (two Ti I and four Ti II) have fairly high intensities of the order of 74, 56, 93, 63, 74 and 99% relative to the main line. These lines are located at 298.31 (Ti II), 334.43 (Ti I), 336.19 (Ti I), 362.58 (Ti
II), 363.34 (Ti II) and 364.35 (Ti II) nm respectively. In addition to these lines, six other lines have lower intensities and are located at 317.64 (Ti II), 319.29 (Ti I), 333.15 (Ti I), 333.99 (Ti I), 337.66 (Ti I) and 372.27 (Ti II) nm; the relative intensities are 21, 22, 27, 28, 26 and 34% respectively. We also noted the presence of eleven spectral lines with an intensity of less than

20% compared with that of the main line. Two of these lines, ($^3F2$) with relative intensities of 08 and 11%, de-excited towards the ground state. These lines are identified with neutral Ti I atoms and are located at 306.52 and 376.44 nm respectively.

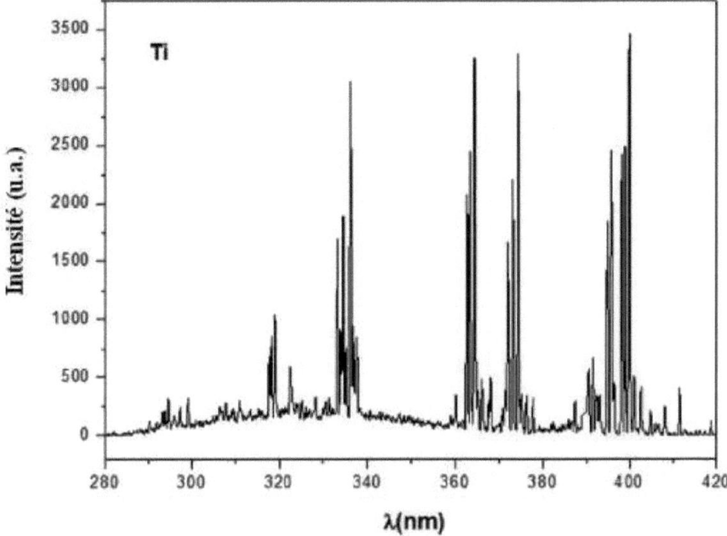

Figure 4-2: Luminescence spectrum of clean titanium

Ti ([Ar] $3d24s^2$ - ^{3}F2)

λo (nm) Observed	λ_L' (nm) Literature [50a,b]	Relative intensities	Transitions [42a,b]	Aki Einstien coefficient (108 s^{-1})[51]
294.70	294.61 Ti II	09	$3d^24d\ e^4G_{11/2}$- $3d^24p\ z^4F_{9/2}$	2.70
298.33	298.31 Ti II	74	$3d^24p\ z^4G_{7/2}$- $3d^24s^2\ a^3F_3$	0.07
297.52	297.46 Ti I	07	$3d^24s5p\ ^3F^\circ_4$- $3d^24s4p\ z^5G^\circ_4$	0.07
306.60	306.52 Ti I	08	$3d^24s5p\ ^3G^\circ_3$- $3d^34s$ hT9	0.21
310.91	310.83 Ti I	09	$3d^24p^2$ sFI- $3d^24s4p\ z^5G^\circ_2$	0.92
317.70	317.64 Ti II	21	$3d^25s\ e^4F_{9/2}$- $3d^24pz$ 4F 2	0.52
319.40	319.29 Ti I	22	$3d^34p\ w^3G_4$- $3d^24s^2\ a^3F_3$	0.85
324.49	324.47 Ti I	08	$3d^24s4p\ v^3D$ 3- $3d^24s^2\ a^3F_4$	0.15
333.26	333.15 Ti I	27	$3d^24s6s\ h^5F_3$- $3d^24s4p\ z^5G^\circ_2$	0.40
334.00	333.99 Ti I	28	.I <2-0 . .<2. <2 . $3d^35p\ ^3G$ 4- $3d^34s\ a^3G_4$	0.05
334.60	334.43 Ti I	56	$3d^24s6s\ h^5F_4$- $3d^24s4p\ z^5G^\circ_4$	1.68
335.33	335.15 Ti I	49	$3d^24s6s\ h^5F_2$- $3d^24s4p\ z^5G^\circ_3$	0.05
336.30	336.19 Ti I	93	2 j2 < л 3 л"-•° 2 i2 < 2 3p 3d 4s4p x G 3- 3d 4s a F3	0.10
337.81	337.66 Ti I	26	$3d^34p\ ^3G^\circ_4$- $3d^34s\ b^3F_3$	0.13
362.60	362.58 Ti II	63	$3d^24p\ z^2S_{3/2}$- $3da^3\ a^2P_{1/2}$	0.29
363.50	363.34 Ti II	74	$3d^24d\ ^4P_{3/2}$- $3d^24p\ z^4S^\circ_{3/2}$	0.30
364.40	364.35 Ti II	99	$3d^24d$ 2G7/2- 3d4s4p $^4D^\circ_{5/2}$	0.49
370.80	370.72 Ti II	07	$3d^24p\ y^2D^\circ_{3/2}$- $3d^3\ b^2D^2_{3/2}$	0.28
372.39	372.27 Ti II	34	2 j2 j 2т7° 2 i2 л 2т-' 3d 4p z F 7/2- 3d 4s a F5/2	0.03
374.30	374.27 Ti II	95	$3d^24p\ y^2D_{5/2}$- $3d^3\ b^2D^2_{5/2}$	0.68
374.96	374.92 Ti I	10	$3d^34d\ ^5G_6$- $3d^24s4p\ z^5F^\circ_5$	0.23
376.50	376.44 Ti I	11	$3d^34p\ y^5G$ 3- $3d^24s^2\ a^3F_2$	0.21
377.71	377.54 Ti II	09	$3d^24p\ z^4F_{7/2}$- $3d^24s\ a^4F_{5/2}$	0.37
402.70	402.62 Ti II	100	$3d^24p\ z^4G_{7/2}$- $3d^24s\ a^2F_{7/2}$	0.03

Table 4-1: Titanium spectral lines and their identification

III.2. 2. Optical emissions from titanium in the presence of oxygen

In order to better understand the influence of oxygen on the depopulation of excited electronic states and consequently on the variation in the light signal, we bombarded titanium in an oxygen atmosphere at a pressure of 10 5 torr. The spectral range scanned extends from 280 to 420 nm, the region in which the majority of spectral lines are found. Figure 4-3 shows the spectrum obtained when bombarding titanium under the same experimental conditions used for clean titanium but at the pressure mentioned above.

In the presence oxygen, we measured the same spectrum as in the absence of this gas, i.e. the same number of Ti I and Ti II lines. However, the intensities of the spectral lines do not change in the same way in the presence oxygen. In fact, a simple comparison of the two spectra of titanium and titanium in the presence oxygen revealed not only an increase but also an invariance in the intensities of the Ti I spectral lines. The intensity of four Ti I lines located at 297.46, 310.83, 333.15 and 334.43 nm remained unaffected.

The intensities of the other nine Ti I lines increased by a factor R of around 7 (Table 4-2). As for the Ti II lines, their intensities decreased by as much as 7%.

The effect of oxygen appears to shift the wavelength of the most intense line, which is now a Ti I line located at wavelength 335.15 nm and is associated with the transition $3d^24s6s\ h^5F_2$- $3d^24s4p\ z^5G^\circ$ 3. In the case of clean titanium, the main line observed is a Ti II line (402.62

nm) intensity has decreased by 50%. Oxygen also has a considerable influence on the intensities of titanium spectral lines. As we shall see in the next paragraph, this influence can be adequately explained within the framework of the electron model for the Ti I lines.

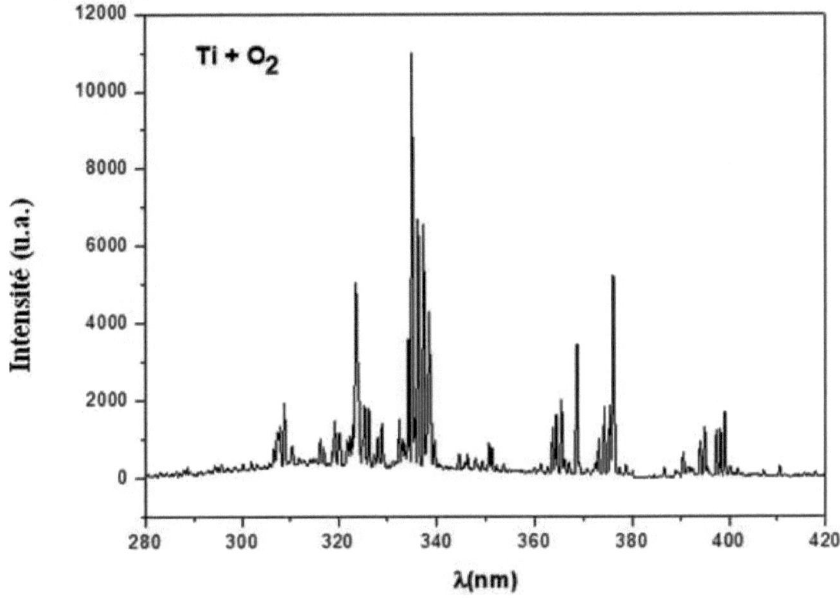

Figure 4-3: Luminescence spectrum of titanium in the presence of oxygen

λ (nm)	Intensity (a.u.)		Spectral terms	$R=\frac{I_{Ti+O_2}}{I_{Ti}}$	D	Energy levels (eV) [50 a,b]			
	Ti (P<10 7 torr)	Ti + O2 (P 10 5 torr)				Superior	Lower	Ev	
297.46 Ti I	245	250	$^{3}F°_{4}$- $^{5}G°_{4}$	1.02	0	6.04	1.88	-0.79	**a**
306.52 Ti I	250	625	$^{3}G°_{3}$- $^{3}F_{2}$	2,50	+	5.47	0.00	-1.36	b
310.83 Ti I	300	325	$^{5}F_{1}$- $^{5}G°_{2}$	1,08	0	5.95	1.97	-0.88	**c**
319.29 Ti I	720	1410	$^{3}G°_{4}$- $^{3}F_{3}$	1,95	+	3.90	0.02	-2.93	d
324.47 Ti I	270	1450	$^{3}D°_{3}$- $^{3}F_{4}$	5,37	+	3.87	0.04	-2.96	e
333.15 Ti I	880	910	$^{5}F_{3}$- $^{5}G°_{2}$	1,03	0	5.69	1.97	-1.14	**f**
333.99 Ti I	910	1370	$^{3}G°_{4}$- $^{3}G_{4}$	1,50	+	5.59	1.88	-1.24	g
334.43 Ti I	1855	1865	$^{5}F_{4}$- $^{5}G°_{4}$	1,00	0	5.70	1.99	-1.13	**h**
335.15 Ti I	1625	10990	$^{5}F_{2}$- $^{5}G°_{3}$	6,76	+	5.58	1.98	-1.25	i
336.19 Ti I	3055	6375	$^{3}G°_{3}$- $^{3}F_{3}$	2,08	+	3.70	0.02	-3.13	J
337.66 Ti I	845	2120	$^{3}G°_{4}$- $^{3}F_{3}$	2,50	+	5.11	2.3	-1.72	k
374.92 Ti I	320	1180	$^{5}G_{6}$- $^{5}F°_{5}$	3,68	+	5.44	2.13	-1.39	l
376.44 Ti I	1351	5202	$^{5}G°_{3}$- $^{3}F_{2}$	3.85	+	3.29	0.00	-3.54	m

Table 4-2: Observed emissions of Ti I lines during bombardment of a sample of clean titanium and titanium in the presence of oxygen.

clean titanium and titanium in the presence of oxygen

λ (nm)	Intensity (a.u.)		Spectral terms	$R=\frac{I_{Ti+O_2}}{I_{Ti}}$	D	Energy levels (eV) [50a,b]		
	Ti (PH)' torr)	Ti + o2 (P 10 5 torr)				higher (eV)	lower (eV)	Ev (eV)

294.61 Ti II	310	285	- 4G11/2 $^4F_{9/2}$	0,91	-	8.08	3.88	-5.55
298.31 Ti II	2420	1215	$^4G°_{7/2}$- 2F5/2	0,50	-	3.68	0.57	-9.90
317.64 Ti II	680	325	4F9/2- $^4F^2_{7/2}$	0,47	-	7.76	3.86	-5.82
362.58 Ti II	2070	285	- 2S1/2 2P1/2	0,13	-	4.64	1.22	-8.94
363.34 Ti II	2445	190	4P3/2- 4S°3/2	0,07	-	8.37	4.96	-5.21
364.35 Ti II	3255	1640	- D 2G7/2 4 5/2	0,50	-	9.93	6.52	-3.65
370.72 Ti II	235	145	D D 2 3/2- 2 23/2	0,61	-	4.91	1.56	-8.67
372.27 Ti II	1120	335	$^2F_{7/2}$- 2F5/2	0,29	-	3.90	0.57	-9.68
374.27 Ti II	3290	1820	2D5/2- 2D25/2	0,55	-	4.89	1.58	-8.69
377.54 Ti II	310	260	4F7/2- 2F5/2	0,83	-	3.85	0.57	-9.73
402.62 Ti II	3457	1697	4G7/2- 2F7/2	0,79	-	3.70	0.61	-9.88

Table 4-3: Observed emissions of Ti II lines during bombardment of a sample clean titanium and titanium in the presence of oxygen

D: Dependency: (+): increase; (-): decrease and (0): insensitive

IV. Discussion and interpretation of results

Tables 4-2 and 4-3 above give the characteristics of all the emissions observed when clean titanium and oxygenated titanium are bombarded by 5keV Kr^+ ion beam. The columns in each of these tables can be described as follows:

Column 1 shows the wavelength positions of the spectral lines.

Column 2 gives the absolute intensities of the various lines observed for clean titanium and oxygenated titanium.

Column 3 represents the spectral terms and the dependence of the spectral lines: (+) means increase in intensity, (-): decrease in intensity and (0): insensitivity.

In the fourth column, we have indicated the intensity ratio R between oxygenated titanium and clean titanium.

The last column of both tables gives the energy positions (in eV) of the upper and lower electronic states involved in the transition under consideration. The value given for the upper state gives the internal energy of the excited atoms relative to the vacuum level, i.e. $E_v = E^* - I^{n+}$, where E^* is the energy of the emitting state and I^{n+} is the single ionisation potential I^+ (Ti) = 6.83 eV and the double ionisation potential I^{++} (Ti) = 13.58 eV [50a,b].

To explain the increase in the intensities of the Ti I lines of neutral atoms within the electron exchange model, it is essential that the emitting states of these lines are located opposite both the metal work function band ($0 < |E_V|$^ V) and the oxide gap ($A_f < |EV| < Af + E_g$). The excited electron can then transit towards the metal and not towards the oxide, which is characterised by a gap. In the presence of the gap, the excited states of the slow particles are depopulated by photon emission and the light signal is amplified. These states are indicated in the table and diagram by the letters (b, d, e, g, i, j, k, l, m).

Figure 4-4 shows the application of the model we used previously [47,12]. It is assumed that the presence of oxygen modifies the energy band structure of the initial target following chemisorption of this gas. The metal is characterised by the work function (Ti) = 4.3 eV [72] and the oxide TiO_2 by the electron affinity A_f which is 1.59 eV [73] and by the gap width Eg which is equal to 2.5 eV [74]. The emitting states of the Ti I lines of neutral atoms are identified by the E_v values. This model suggests the formation a new titanium oxide characterised by a band gap. We estimate a width of 2.40 eV for this band. This value is defined by the difference between the two energy states noted m ($|E_v|$ = 3.54 eV) and **f** (|EJ = 1.14 eV). This value is in good agreement with the value determined by Mobray et al [74]. The electronic affinity can estimated from the lines intensity remains insensitive. The associated states, which we have denoted by the letters **a**, **c**, **f** and **h,** are located opposite both the empty part of the conduction band of the metal and the conduction band of the oxide. In this case, the excited state can exchange its electron with both the metal and the oxide. With these considerations in mind, we suggest a value of 1.14 eV; this value corresponds to the highest of the energies| EJ associated with the **f** emitting state.

By adopting this diagram (Fig.4-4), which assumes the formation of an oxide film, the behaviour of all the Ti I lines can be adequately explained when we go from clean metal to oxygenated metal.

However, the eleven Ti II lines associated with simply charged ions and which show a negative dependence on oxygen do not follow the behaviour expected from the application of the model because, according to this model, the excited states of these lines can capture an electron from the Fermi bath of the metal and not the oxide according to the *reaction*$Ti^{++} + e''$ Ti^{+} *. To do this, these emitter states must be located opposite the gap but also opposite the occupied states of the conduction band, so that they are populated by the metal and not the oxide. It is therefore likely that other mechanisms are involved in ion formation. One example is the chemical model used for the emission of secondary ions from oxides or metals bombarded in the presence of oxygen or by O_2 ions^{+} . According to this chemical model, the particle is expelled from a position where it exists naturally as an ion in its fundamental state and without electrons neutralising it at the exit [75].

As we recorded a decrease in the intensities of the Ti II lines in an oxygen atmosphere, we can assume that these ions are formed in their ground state.

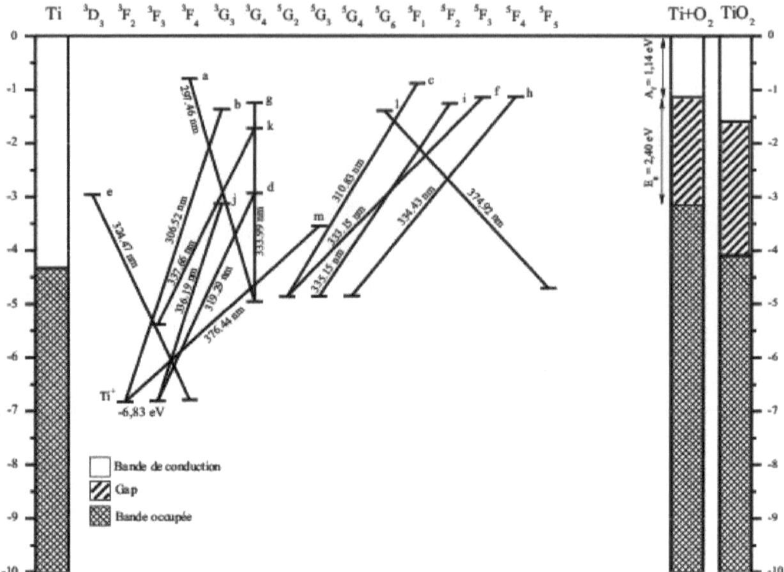

Figure 4.4: Diagram of the energy levels of the metal Ti, the oxide TiO_2, Ti+O2 oxide and the emitting states of the Ti I lines.

III.2 Continuous radiation

In addition to spectral lines, ion bombardment of certain targets gives rise to continuum radiation. There are two types of continuum: The first type is the continuum emitted by the solid itself (ionoluminescence). This type depends both on the nature and composition of the bombarded target and on the energy and angle of incidence of the projectile. Sharodi et al [44] observed a continuum during bombardment of a beryllium target by He ions$^+$ at 15 keV and interpreted its origin as the result of electron-hole recombination. The second type is the continuum emitted by particles outside the target. The sources proposed for this type of continuum are $MO^{(*)}$) excited oxide molecules, polyatomic metallic species, braking radiation (Bremsstrahlung) and implanted or adsorbed molecular species (O_2, H_2, Al H, etc.).

In our work, the luminescence spectrum observed during bombardment of titanium samples both in the absence and presence of oxygen by Kr^+ ions with a kinetic energy of 5 keV essentially comprises a continuous background on which a multitude of spectral lines are superimposed. The continuum we observed extends from 275.00 to 400.00 nm and is centred at 325.00 nm. This continuum was exalted when the titanium was bombarded in an oxygen atmosphere and its intensity doubled (Fig.4-4). These results complement previous work [76] which shows that the most likely source of continua is related to the sputtered particles and therefore the radiation occurs outside the target rather than in the solid. These results are in good agreement with those of Kiyan et al [64] who observed a continuum when bombarding titanium with He^+ , Ne^+ and Ar^+ ions with an energy of 30 keV. Ait El Fqih et al [17] observed continuous radiation between 250 and 500 nm when bombarding a vanadium target under the same experimental conditions. The authors assumed that this radiation came from the collective excitations of the electrons in the d sublayer of the sputtered atoms of the metal.

This observation was detailed and confirmed by El Boujlaidi et al [14] who studied the luminescence spectra of some transition metals and their oxides when bombarded by Kr^{+} at 5 keV. They showed that the spectral range and intensity of the continua observed appear to be related to the electronic structure of the element, more precisely to the number of electrons in the d sublayer. The continuum is much more pronounced for vanadium (3er on 3d) than for titanium (2e~ on 3d), whereas it is totally absent in the case of chromium (5e on 3d), iron (6e on 3d), nickel *(7er* on 3d) and copper (10e on 3d), as well as for light metals (Mg and Al).
Further experiments are required to elucidate the mechanism(s) responsible for the emission of continuous radiation.
The table below summarises the observations made in our laboratory.

Targets studied	Electronic configuration	Continuous radiation
Mg	[Ne] $3s^{2}$	No
Al	[Ne] 3s23p1	No
If	[Ne] 3s23p2	No
Ti	[Ar] 3d24s2	**Yes**
V	[Ar] $3d^{3}$4s2	**Yes**
Cr	[Ar] 3d54s1	No
Fe	[Ar] 3d64s2	No
Ni	[Ar] 3d84s2	No
Cu	[Ar] 3d104s1	No
Visit	[Xe] 4f1^{4}5d^{10}6s1	No

Table 4-4: Continuous radiation from the various targets studied

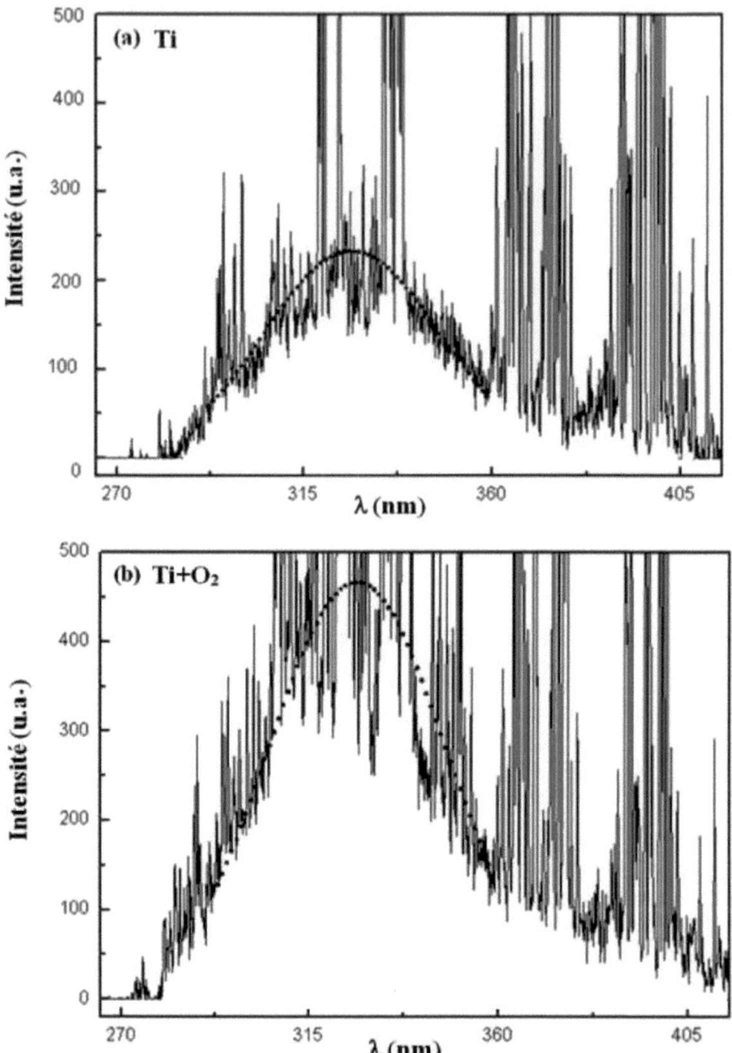

Figure 4-4: Emission spectra showing the continuum zone of samples of: a) clean titanium, b) titanium in the presence of oxygen.

Conclusion

We have demonstrated the optical emission during bombardment of clean titanium samples and titanium in the presence of oxygen by Kr ions^{+} of 5 keV at an angle of incidence of 65° . This optical radiation appears in the form of discrete lines superimposed on a continuum. The lines are identified as Ti I associated with neutral atoms and Ti II from simply charged ions. Our results show that the presence of oxygen during the bombardment of the target under study causes the intensities of the majority of the Ti I lines to increase, while other lines remain insensitive to the presence of this gas. However, the intensity of the Ti II ion lines

decreased.
The electron exchange model accounts for the entire behaviour of Ti I lines and there is a strong presumption that, in the presence of oxygen, a structure is formed at the surface with energy band pattern intermediate between that of the metal and that of the corresponding oxide. For the Ti II lines, other mechanisms are involved, such as chemical emission leading to the formation of ions in the ground state.
The continuous radiation we observed is an optical emission from a sputtered particle. It is due to collective excitations of the electrons in the d-sublayer of the sputtered atoms of the target. The tendency of a metal target to emit continuous radiation during bombardment with ion beam can be closely correlated with the number of electrons in the d-sublayer. It should be noted that this continuous radiation is only observed in the case of transition metals for which the d-sublayer is less than half filled.
However, the origin of this continuous background remains controversial. Some of it must obviously be attributed to molecular emissions of diatomic or even polyatomic oxygen compounds from the bombarded target.

General conclusion

The work presented in this thesis concerns the luminescence of nickel, chromium, titanium and NiO and Cr2O3 oxide sputtering products subjected to bombardment a 5 keV Kr^+ ion beam. The experiments were carried out using the SIPS spectrometer installed in our SIAM laboratory and described in detail in Chapter 2. The targets were etched under a residual gas pressure of better than 10 7' torr and in the presence of oxygen at a pressure of 10 5 torr.

The luminescence spectra measured extend between 190 and 650 nm with a resolution of 0.2 nm, but only over a small spectral region that appears to be of interest. The optical radiation observed in the case of nickel and chromium is in the form of discrete lines attributed to the de-excitation of the Ni I and Cr I neutral atoms. In the spectral region explored, no lines associated with the deexcitation of the ions were observed, nor was any continuous radiation detected. As for the titanium target, the luminescence spectrum obtained contains discrete Ti I and Ti II lines that are superimposed on a continuum. The continuum extends from 275 to 400 nm and is centred at 325 nm. For the three targets studied, we found no optical emission associated with the krypton gas used for bombardment.

In the presence of oxygen, the intensities of the spectral lines change to a greater or lesser extent depending on the transition involved. This change consists of an increase, decrease or invariance in intensity. We have adequately explained this behaviour in the context of the electron exchange model, which assumes the formation of an oxide layer following chemisorption of oxygen at the surface of the target. The formation of this layer is in strong competition with the erosive effect of the bombardment beam and, as a result, the transformation into oxide is not complete. We considered that the behaviour of optical emissions with this gas is governed by a situation in which a new system is formed with a composition and band structure intermediate between those of the metal and those of the native oxide. This system is a kind of metal-oxygen 'alloy' whose strechiometry is different from that of the oxide, but characterised by an energy band gap. The presence of the gap blocks non-radiative de-excitation processes and therefore allows the excited states of the particles to be depopulated by photon emission. In the case of oxygenated metals, this mechanism leads to an increase in the light signal from the lines associated with neutral atoms. In addition to this mechanism, the radiative dissociation of pulverised molecules can give rise to excited atoms that are deactivated far from the surface by photon emission.

The presence of oxygen also causes a decrease in the intensity of the lines due to the de-excitation of ions (ionic lines). The recommended model does not account for this decrease and it is very likely that other mechanisms are involved in ion de-excitation. We have suggested the mechanism of chemical emission, which leads to the formation of ions in the ground state and therefore to ions that do not radiate.

In the case a clean target (bombardment carried out at a base pressure of better than 10 7' torr), the light observed comes mainly from fast particles (-107 cm/s) which leave the surface directly in excited electronic states. These particles deactivate far from the surface at distances where their excitation levels are not disturbed by the presence of the solid. Light can also come from slow particles ejected in a given state of excitation or charge. However, the contribution of these particles is largely governed by the efficiency of the electron exchange processes (neutralisation or ionisation by resonance and de-excitation or Auger neutralisation) with the metal's conduction band.

Continuous radiation (continuum) was observed under our experimental conditions in the case of a titanium target and not nickel or chromium. This radiation appears to have a more complex origin than that described in the literature. In fact, this radiation does not come from sputtered molecular species such as MO molecules (where M is a metal atom) since, in the presence of oxygen, this emission is completely absent in a large number of metal targets. We have suggested that this radiation is linked to the collective de-excitation of electrons in the 3d sublayer. These results and others from our laboratory show that continuum emission exists when the number of electrons in the 3d sublayer is less than 5 (3d filled less than half) and that it is due to collective de-excitation of these electrons.

With this work, we believe we have contributed to the study of the light emitted during the sputtering of nickel, chromium and titanium following ion bombardment. This work, which is necessarily incomplete, needs to be followed up by other studies such as the post-excitation of the sputtered particles by a laser or an electron beam of appropriate energy. This type of study could elucidate the mechanisms responsible for the change in the yield of photons emitted during ion beam-induced sputtering of a solid.

References

[1] C. K. Singh, S. Ilango, S. Dash, A.K. Tyagi, "Secondary ion mass spectrometry based depth profiling of Mo/Si interfaces with different microcrystalline structure", Mater. Chem. and Phys. 173 (2016) 475-481.

[2] H. Oechsner, "Secondary neutral mass spectrometry (SNMS)- recent methodical progress and applications to fundamental studies in particle/surface interaction", International Journal of Mass Spectrometry and Ion Processes, 143 (1995) 271-282.

[3] A. O. Juma, P. Pistor, S. Fengler, T. Dittrich, E. Wendler, "Copper diffusion in thin In_2S_3 layers investigated by rutherford backscattering spectroscopy (RBS)", Thin Solid Films, 520 (2012) 6740-6743.

[4] C. T. Campbell, T. E. James, "Ion scattering spectroscopy intensities for supported nanoparticles: The hemispherical cap model", Surface Science, 641 (2015) 166-169.

[5] N. Vaidyanathan, M. Houalla, D. M. Hercules, "Surface Coverage of Catalysts WO_3/ZrO_2 Measured by Ion Scattering Spectroscopy and Low Temperature Adsorption", Surf. Interface Anal. 26 (1998) 415-419 .

[6] N. Sharifzadeh, O. Kakuee, S. Mohammadi, "Measurements of proton induced γ-ray emission cross-sections on Mg from 1.0 to 3.0 MeV", Nucl. Inst. Meth. B 372 (2016) 109113.

[7] M. Suchanska, "Ion-induced photon emission of metal", Progress in Surface. Science, 54 (1997) 165-209.

[8] A. Kaddouri, I. Ashraf, M. Ait El Fqih, H. Targaoui, A. El Boujlaidi, K. Berrada, "Photon emission from clean and oxygenated Si and SiO_2 surfaces bombarded by 5 keV krypton ions", Appl. Surf. Sci. 256 (2009) 116-119.

[9] P.G. Fournier, J. Fournier, B. Bellaoui, "Optical emissions from sputtering species formed by the fast ion bombardment of clean and oxygenated Mg surfaces", Nuc. Instr. Meth. B 67 (1992) 604-609.

[10] C.S. Lee, Z.M. Lin, Z.M. Yen, C.H. Lin, "Light emission from oxygen covered Al and Cu surfaces", Nucl. Inst. Meth. B 190 (2002) 141-145.

[11] K. Hammoum, T. El Asri, A. Chahboune, H. Aouchiche and A. Kaddouri, "Optical emission of products sputtered from Fe, Fe_2O_3 and Fe_3O4 powders", Eur. Phys. J. D, 61 (2011) 469-474.

[12] A. El Boujlaidi, K. Hammoum, L. Jadoual, R. Jourdani, M. Ait El Fqih, H. Aouchiche, A. Kaddouri, "Photon emission produced by Kr^+ ions bombardment of Cr and Cr_2O_3 targets", Nucl. Instrum. Meth. B 343 (2015) 158-162.

[13] T. El Asri, M. Raissi, S.Vizzini, A. El Maachi, E. L. Ameziane, F. Arnaud d'Avitaya, J. L. Lazzari, C. Coudreau, H. Oughaddou, B. Aufray, A. Kaddouri, "Inter-diffusion of cobalt and silicon through an ultra thin aluminum oxide layer", Appl. Surf. Sci. 256 (2010) 27312734.

[14] A. El Boujlaidi, M. Ait El Fqih, K. Hammoum, H. Aouchiche, A. Kaddouri, "Continuum radiation emitted from transition metals under ion bombardment", Eur. Phys. J. D 66 (2012) 1-6.

[15] G. Louarn, K. Berrada, N. Errien, M. Ait El Fqih, G. Froyer, A. Kaddouri, "Vibrational and optical emission spectoscopies of porous and oxydized porous silicon", Phys. Chem. News, 21 (2005) 6-11.

[16] M. Ait El Fqih, L. Jadoual, A. El Boujlaidi, R. Jourdani and A. Kaddouri, "Comparative study of the angular distribution of Cu and Be atoms sputtered from a CuBe Alloy target

under Kr^+ Bombardment", J. of Active and Passive Electronic Devices, 10 (2016) 273-281.
[17] M. Ait El Fqih, A. El Boujlaidi, R. Jourdani, A. Kaddouri, "On the validity of the electron transfer model in photon emission from ion bombarded vanadium surfaces", Eur. Phys. J. D 63 (2011) 97-102.
[18] P. Sigmund, "Theory of sputtering. Sputtering yield of amorphous and polycrystalline targets", Phys. Rev. 184 (1969) 383-416.
[19] A. Benninghoven, R.J. Colton, D. S. Simons, H.W. Werner, "Secondary ion mass spectroscopy SIMS V", Springer Series in chemical Physics, 44 (Springer, Berlin-Heidelberg-New York- Tokyo, 1986).
[20] J. F. Ziegler, "SRIM-2003", Nucl. Inst. Meth. B 219 (2004) 1027-1036.
[21] Christophe Verdeil. "Etude de la pulvérisation et de l'émission de la matière sous bombardement Cs^+ ", thesis, Institut National Polytechnique de Lorraine (2008).
[22] P. Sigmund, "Fundamental processes in sputtering of atoms and molecules (SPUT92)", Symposium on the occasion of the 250th Anniversary of the Royal Danish Academy of Sciences and Letters Copenhagen (1992) 7-659.
[23] G. Carter, J. S. Colligon, "Ion bombardment of solids", Heinemarm Educational Books Ltd, London, 12 (1968) 120-446.
[24] M. Nastasi, J.W Mayer, J.K Hirvonen, "Ion-solid interaction", Fund. and Appl., Chap. 9, Cambridge Solid State Science (1996).
[25] A. Karen, Y. Nakagawa, M. Hatada, K. Okuno, F. Soeda, A. Ishitani, "Quantitative investigation of the O_{2+} induced topography of GaAs and other III-V semiconductors: An STM study of the ripple formation and suppression of the secondary ion yield change by sample rotation" , Surf. Inter. Anal, 23 (1995) 506-513.
[26] H. Oechsner, "Sputtering - a review of some recent experimental and theoretical aspects", Appl. Phys. 8 (1975) 185-198.
[27] A. E. Morgan, H. A. M. Grefte, N. Warmoltz, H. W. Werner and H. J. Tolle, "The influence of bombardment conditions upon the sputtering and secondary ion yields of silicon", Appl. Surf. Sci. 7 (1981) 372-392.
[28] G. Betz, W. Husinsky, "Sputtering of metal targets under increased oxygen partial pressure", Nucl. Inst. Meth. B 13 (1986) 343-347.
[29] E. W. Thomas, "Formation of excited states by ion impact on surfaces", Progress in Surface Science, 10 (1980) 383-489.
[30] S. Kunitomo, M. Ono, C. F. Li, R. Shimizu, "Energy distributions of atoms sputtered from Cu- Pt Alloy measured by nonresonant multiphoton ionization sputtered neutrals mass spectrometry", Jpn. J. Appl Phys, 32 (1993) 3991-3994.
[31] P. Sigmund, "Interaction of Particles and Radiation with Matter", Nucl. Inst. Meth. B 155 (1999) 337-338.
[32] H. L. Bay, W. Berres, "Anisotropy of collision cascades by light-ion irradiation of zirconium" Nucl. Instrum. Meth. B 2 (1984) 606-609.
[33] Ait El Fqih, A. el Boujlaidi, A. Kaddouri, "Angular distributions of sputtered beryllium at normal incidence", Insta. J. Phys. 2 (2012) 89-94.
[34] V.I. Shulga, "Angular distribution of atoms sputtered from amorphous and polycrystalline targets", Nucl. Inst. Meth. B 164 (2000) 733-747.
[35] H. Snoek, W. F. van der Weg, R. Geballe, P. K. Rol, "Inelastic energy losses in collisions of 40-90 keV Ar_+ ions with solid and gaseous Cu targets", Physica, 35 (1967) 1-18.
[36] W. F. van der Weg and D. J. Bierma, "Excitation of Cu atoms by Ar ions and subsequent

radiation less deexcitation of scattered particles near a Cu surface", Physica, 44 (1969) 206218.
[37]P. J. Martin, A. R. Bably, R. J. MacDonald, N. H. Tolk, G. J. Clark, J. C. Kelly "Deexcitation processes near the surface ion bombarded SiO, and Si", Surface. Science, 60 (1976) 349-364.
[38]C. W. White and N. H. Tolk, "Optical Radiation from Low-Energy Ion-Surface Collisions", Phys. Rev. Letters, 26 (1971) 486-489.
[39]Terzic and Perovic, "Spectral analysis of light emitted from metallic targets bombarded by high energy ions", Surface Science, 21 (1970) 86-98.
[40]A. Kaddouri, "Spectroscopie d'émission des produits de pulvérisation de solides soumis à un bombardment ionique", thesis, Université Paris- Sud, N d'ordre 3529 (1989).
[41]Chin Shuang Lee, Tzi Ming Yen, Jing Hui Lin, "Light emission from an oxygen covered copper surface by ion bombardment", Surface Science, 488 (2001) 379-386.
[42]Ait El Fqih, A. Fake, K. Berrada, A. Kaddouri, "Effet de l'oxygène sur les radiations optiques émises lors de la pulvérisation", Afrique Science, 3 (2008) 339-356.
[43]J. N. Murrell, S. D. Bosanac, "Introduction to the theory of atomic and molecular collisions", Chemical Society Reviews, (1992) 17-28.
[44]I. S. Sharodi, Yu. A. Bandurin, S. S. Pop, "Ion photon emission during ion bombardment of beryllium surface", Nucl. Instr. Meth. B 193 (2002) 699-704.
[45]A. Muller, A. Benninghoven, "Investigation of surface reactions by the static method of secondary ion mass spectroscopy III. The oxidation of vanadium, niobium and tantalum in the monolayer range", Surface Science 39 (1973) 427-436
[46]H. S. Ousmane, T. Elasri, E. Ech-chamikh, M. Azizan, A. Essafti, Y. Ijdiyaou A. Kaddouri, "Optical Emission Spectroscopy Study Of RF Sputtered A-C/WOX Interface", M. J. Condensed Matter, 12 (2010) 284-286.
[47]L. Jadoual, A. El Boujlaidi, M. Ait El Fqih, A. Aamouche, K. Kaddouri, "Optical emission from ion-bombarded nickel and nickel oxide", Spectroscopy Letters, 47 (2014) 363366.
[48]A. Tsukazaki, H. Saito, K. Tamura, M. Ohtani, H. Koinuma, M. Sumiya, S. Fuke, T. Fukumura, M. Kawasaki, "Systematic examination of carrier polarity in composition spread ZnO thin films codoped with Ga and N", Appl. Phys. Lett. 81 (2002) 235-237.
[49]J. C. Pivin, C. Roques-Carmes, G. Slodzian, "Layers formed on Fe-Ni, Fe-Cr, Ni-Cr alloys under ion bombardment in the presence of oxygen", International Journal of Mass spectrometry and ion physics, 31 (1979) 311-332.
[50]a) Radzig, A. A. Smirnov, B. M. Reference Data on Atoms, Molecules, and Ions. Springer-Verlag: Berlin (1985).
b) Linde, D. R. CRC Handbook of Chemistry and Physics, 83rd ed. CRC Press: Washington DC, E 279 (2002).
[51]R. Striganov, N. S. Sventitskii, "Tables of spectral lines of neutral and ionized atoms", IFI/Plenum Data Corporation 1968.
[52]W.F.Van der weg and E. Lugujjo, "atomic collisions in solids", Ed. S. Datz. B. R. Appleton, C. D. Moak, Plenum Press, New York, 2 (1975) 511-735.
[53]Michaelson, B. Herbert, "The work function of elements and its periodicity", J. Appl. Phys. 48 (1977) 4729-4733.
[54]W. Hongbin, W. Lai-Sheng. "A study of nickel monoxide (NiO), nickel dioxide (ONiO), and Ni (O2) complex by anion photoelectron spectroscopy",J. Chem. Phys. 107 (1997) 1621.

[55]G. J. M. Janssen, W. C. Neiuwpoort, "Band gap in NiO a cluster study", Phys. Rev. B 38 (1988) 3449-3458.
[56]] C. C. Shull, M. K. Wilkmson, "Neutron Diffraction Studies of Various Transition Elements", Rev. Mod. Phys. 25 (1953) 100-107.
[57]...] David R. Lide, "CRC Handbook of Chemistry and Physics", CRC Press/Taylor and Francis 90th ed. 2009.
[58]Y. Mito, M. Ueda, T. Ohtsuka, "Photo-luminescence from passive oxide films on nickel and chromium by photo-excitation of UV light", corrosion science, 51 (2009) 1540-1544.
[59]R. H. Misho, W. A. Murad, G. H. Fattahallah, "Preparation and optical properties of thin films of CrO_3 and Cr_2O_3 prepared by the method of chemical spray pyrolysis", Thin Solid Film, 169 (1989) 235-239.
[60]M. F. Al-Kuhaili, S. M. A. Durrani, "Optical properties of chromium oxide thin films deposited by electron-beam evaporation", Optical Materials, 29 (2007) 709-713.
[61]P. Hones, M. Diserens and F. levy, "Characterization of sputter-deposited chromium oxide thin films", Surf. and Coat. Tech, 120 (1999) 277-283.
[62]M. Julkarnain, J. Hossain, K. S. Sharif, K. A. Khan, "Temperature effect on the electrical properties of chromium oxide (Cr2O3) thin films", J. of Optoelec. And Adv. Mater, 13 (2011) 485-490.
[63]K. Wagatsuma, "Wavelength table of chromium emission lines in argon glow discharge optical emission spectrometry", J. Anal. Chem. 367 (2000) 414-415.
[64]R. J. Mac Donal, P. J. Martin, "A study of the interaction of oxygen with chromium using ion bombardment induced photon and secondary ion emission", Surface Science, 67 (1977) 237-250.
[65]T. S. Kiyan, V. V. Gritsyna, Ya. M. Fogel, "On the continuous spectrum emitted by particles ejected from the surface of solid targets by an ion beam", Nucl. Inst. Meth. B 132, (1976) 415-417.
[66]Chunhong Li, Feng Pan, Xiujin Wang, Lijuan Wang, He Wang, Haibo Wang, Donghang Yan, "Effect of the work function of gate electrode on hysteresis characteristics of organic thin-film transistors with Ta2O5/polymer as gate insulator", organic Electronics, 10 (2009) 948-953.
[67]Y. Sakisaka, H. Kato, M. Onchi, "Oxygen chemisorption and initial oxidation of Cr (110)", Surface Science, 120 (1982) 150-170.
[68]M. Parker, B. Stec, "An alternative approach to plating of titanium and Ti alloys using carbon foam substrate", Metal Finishing, 110 (2012) 19-22.
[69]J. M-Valtierra, C. F-Reyes, M. O-Morales, "Phase transformation in semi-transparent TiO2 films irradiated with co2 laser", Materials Letters, 66 (2012) 172-175.
[70]M. Buchalska, M. Surówka, J. Hämäläinen, T. Iivonen, M. Leskelä "Photocatalytic activity of TiO_2 films on Si support prepared by atomic layer deposition", Catalysis Today, 252 (2015) 14-19.
[71]T.E. Parker, Roger Kelly, "Ion-impact chemistry in the titanium-oxygen system (studies on bombardment-enhanced conductivity-III)", Journal of Physics and Chemistry of Solids, 36 (1975) 377-385.
[72]K.V. Korshunov , M.V. Tsarev, V.V. Mokrushin, A.M. Shapovalov, E.V. Zabavin "Application of impedance spectroscopy to study oxidized powders of titanium hydride", Journal of Alloys and Compounds, 645 (2015) 140-143.
[73]A. Imanishi, E. Tsuji, Y. Nakato, "Dependence of the work function of TiO_2 (rutile) on

crystal faces, studied by a scanning Auger microprobe", J. Phys. Chem. C, 111 (2007) 21282132.
[74]H. Wu, L.-S. Wang, "Titanium oxide clusters", J. Chem. Phys. 107 (1997) 8221-8228.
[75]D. J. Mowbray, J. I. Martinez, J. M. Garcia Lastra, K. S. Thygesen, K. W. Jacobsen, "Stability and electronic properties of TiO2 nanostructures with and without B and N doping", J. Phys. Chem. C, 113 (2009) 12301-12308.
[76]G. Blaise and G. Slodzian, "Effets comparés de l'oxygène sur l'émission ionique et le potentiel de surface des métaux", Surface Science, 40 (1973) 708-714.
[77]O. Varenne, "Analyse des produits de pulvérisation de métaux et d'oxydes par spectroscopie optique", thesis, Université Paris-Sud Orsay (2000).

Appendix

Angular distribution

This study focuses on the angular distribution of the sputtered product measured during bombardment of a target by ion beam. To do this, we used part of the SIPS spectrometer described in detail in Chapter 2 of this thesis. We used the ion gun and sample chamber of this spectrometer. We had to place a sheet of Mylar in this chamber surrounding the target to be bombarded. The choice of Mylar sheet was justified by the fact that it has a bonding coefficient close to unity[*1] and has a low outgassing rate during pumping. In addition, when used in a solvent, the particles can be easily removed. The experimental results obtained are compared with a simulation using the SRIM[**] programme[2] and then compared with a law giving the angular distribution.

In the first stage, we used copper as the bombarded target. This metal was placed in the centre of a cylindrical support inside the adjustable sample chamber. We then folded the Mylar sheet into the cylindrical shape of the enclosure after drilling a hole to allow the ion beam to pass through. After pumping and switching on the source to produce the ion beam, the copper target was bombarded for 75 hours at normal incidence (0° relative to the normal of the sample surface) by a beam of krypton ions at an energy of 5 keV. The average ion current measured on the sample is practically 1.25 μA.

When the ion beam strikes a target, the pressure increases slightly to 10 7' torr. This increase in pressure is due to the presence of the flow of ions incident on Kr^+ and to the sputtered particles resulting from the impact of the ion beam with the walls of the diaphragm located just before the sample chamber. Inside the chamber, the Mylar sheet takes on a cylindrical shape of radius R, the axis of which passes through the point of impact. Figure 1 shows a diagram of the Mylar cylinder with the following parameters: a is the angle of attack relative to the normal to the surface of the target sample, Θ is the angle between the normal to the sample and the velocity vector of the ejected particle. A point with coordinates (x,y) on this sheet is thus viewed at an angle λ with respect to the plane of incidence such that $\lambda = \text{artan}(\frac{y}{R})$ with R the cylinder radius and μ the angle with respect to the normal in the plane of incidence, i.e. : $\mu = \frac{x}{R}$.

[1] P.-G. Fournier, O. Varenne, J. Baudon, A. Nourtier, T.R. Govers, "Experiment combining ion-beam sputtering and quantitative ICP-OES analysis: angular distributions and total yield of titanium sputtered by 5 keV krypton ions", Appl. Surf. Sci. 225 (2004) 135-143.

[2] . F. Ziegler, "SRIM-2003", Nucl. Inst. Meth. B 219 (2004) 1027-1036.

The normal to the surface is defined by μ= λ= α= O
The element of the solid angle is given by :

$$d\Omega = \cos^3(\lambda \frac{dS}{R^2})$$

dS being the surface element of the Mylar sheet.

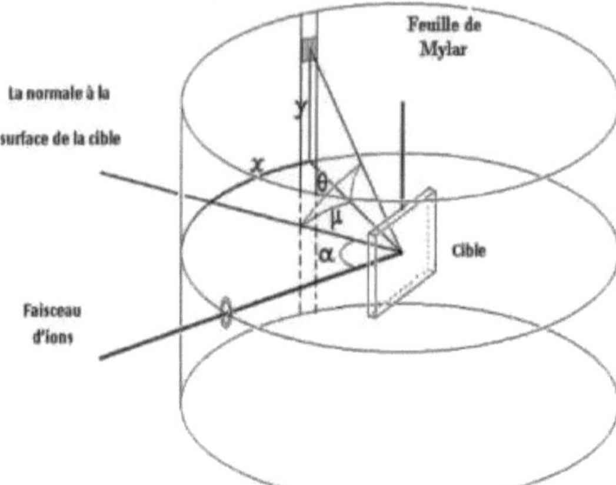

Figure 1: Experimental set-up for collecting spray particles

Figure 2 shows an image of the Mylar sheet after deposition of the spraying product. Around the aperture (through which the beam passed) is a black elliptical spot with a long axis b=68 mm and a short axis a=60 mm. This spot clearly shows the deposition of particles from the target stuck to the Mylar sheet. Also, very close to the aperture, there is a very dark black colour, which is very probably linked to the large number of particles deposited in this region of the Mylar sheet.

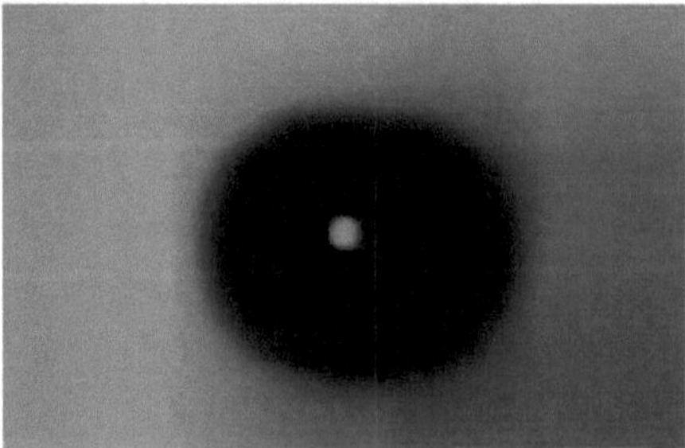

Figure 2: Deposition on the Mylar sheet

Briefly, the results of this work are :

> The elliptical shape obtained on the collector substrate confirms that obtained by simulation using the SRIM software combined with the ANGULAR software.

> The crater caused by the ion beam on the target has an elliptical shape. This shape is probably due to the fact that the direction of the beam is not perfectly perpendicular to the surface of the sample.

> By measuring the dimensions of the crater, we were able to determine the volume caused by the ion beam and subsequently the quantity of material ejected.

The second part of this study presents the results of the angular distributions obtained in the case of ion bombardment of the CuBe alloy with 2% beryllium. These distributions were obtained at angles of 0, 30, 70 and 80° to the normal to the surface of the sample. The target was bombarded a period of about 59 hours and the intensity of the ion current fluctuated very slightly around 2.1 μ A.

To measure the quantity of material deposited on the collector substrate, we cut it according to a very precise pattern[***]. [3]All the results obtained are detailed in the attached publication. The most important points were

- The law followed is a sur-cosine law (of the angle between the direction of the ejected particle and the normal to the surface).
- The presence of significant roughness induced by ionic etching of the surface.
- The shape of the angular distributions tends to decrease more slowly for larger angles of incidence, both experimentally and by simulation.

[3] M. Ait El Fqih, L. Jadoual, A. El Boujlaidi, R. Jourdani and A. Kaddouri, "Comparative study of the angular distribution of Cu and Be atoms sputtered from a CuBe Alloy target under Kr Bombardment", J. of Active and Passive Electronic Devices, 10 (2015)
273-281.

Printed by Books on Demand GmbH, Norderstedt / Germany

Printed by Books on Demand GmbH, Norderstedt / Germany